PENGUIN BOOKS

PURE, WHITE, AND DEADLY

John Yudkin (1910–1995) was a British physiologist and nutritionist. He became internationally famous with his book *Pure, White, and Deadly*, first published in 1972, and was one of the first scientists to claim that sugar was a major cause of obesity and heart disease.

Robert H. Lustig, M.D., is the *New York Times* bestselling author of *Fat Chance: Beating the Odds Against Sugar, Processed Food, Obesity, and Disease*. His YouTube video "Sugar: The Bitter Truth" has been viewed more than three million times. An internationally renowned pediatric endocrinologist at the University of California, San Francisco, Lustig has spent more than fifteen years treating childhood obesity and studying the effects of sugar on the central nervous system, metabolism, and disease.

John Yudkin

Pure, White, and Deadly

How Sugar Is Killing Us and

What We Can Do to Stop It

Introduction by Robert H. Lustig, M.D.

Previously published as *Sweet and Dangerous*

PENGUIN BOOKS

PENGUIN BOOKS

Published by the Penguin Group
Penguin Group (USA), 375 Hudson Street,
New York, New York 10014, USA

USA I Canada I UK I Ireland I Australia I New Zealand I India I South Africa I China
Penguin Books Ltd. Registered Offices: 80 Strand, London WC2R 0RL, England
For more information about the Penguin Group visit penguin.com

First published as *Sweet and Dangerous* in the USA by Peter H. Wyden 1972
First published under the present title in Great Britain by Davis-Poynter 1972
Revised and expanded edition first published in Great Britain by Viking 1986
First published in Penguin Books (UK) 1988
Edition with an introduction by Rober Lustig published 2012
Published in Penguin Books (USA) 2013

LIBRARY OF CONGRESS CATALOGING IN PUBLICATION DATA
Yudkin, John, 1910–1995.
   [Sweet and Dangerous]
   Pure, white, and deadly : how sugar is killing us and what we can do to stop it /
John Yudkin ; introduction by Robert H. Lustig, M.D.
      pages cm
   "Previously published as Sweet and dangerous."
Includes bibliographical references and index.
   ISBN 978-0-14-312518-1 (pbk.)
   1. Carbohydrates, Refined—Physiological effect.   I. Title.
   QP701.Y83 2013
   612.3'96—dc13   2013025742

Printed in the United States of America
ScoutAutomatedPrintCode

Set in Minion Pro Regular
Designed by Alissa Amell

For Benjamin, Ruth, and Daniel

Contents

Acknowledgments

Much of the experimental work that I shall cite here was carried out in the Department of Nutrition at Queen Elizabeth College. I have been most fortunate in having had, over several years, many colleagues and research students who have contributed greatly to the ideas and to the hard work involved in the slow—the enormously slow—unraveling of some of the problems that we have tackled. Without their collaboration many of the facts I quote would not have been known.

Finally, I must say here how grateful I am to the many firms in the food and pharmaceutical industry that for 25 years have given me such constant generous support in the building up and maintenance of the Department of Nutrition. For many of them, the results of our research were often not at all in their interests, yet it was largely with their help that we were able to work on those problems that, to me, seemed of such importance.

Prophecy and Propaganda

Introduction to the 2012 edition by
Robert H. Lustig, M.D.

Everything old is new again. Take fashion, for example: bell-bottoms, culottes, miniskirts, wedge heels, thin ties, and fancy lingerie are back. A silent film won the Oscar for Best Picture in 2012. The bubblegum rock band ABBA and swing-dancing are in vogue again. Specialty cocktails are making a comeback: martinis are the rage, and now there are eighty varieties. Even phonographs and vinyl LPs have a new following.

Ideas come and go as well. Someone is always on the cutting edge. The argument seems inescapable. It gains a following, sometimes a bit too zealous a following. Then it falls out of fashion, due sometimes to philosophy, sometimes to experience, sometimes to competing world events, and sometimes to dark forces attempting to maintain the status quo for their own purposes.

But science should be based in fact, not fashion. And policy should be based on science. Facts shouldn't change. And indeed, they don't. But their interpretation does. Consider the idea that inflammation causes heart disease. First espoused in the late 1800s after the invention of aspirin by Bayer, this idea was relegated to the dustbin of medical science in favor of the cholesterol hypothesis, which reigned for the second half of the twentieth century. But over the last decade, the "inflammation hypothesis" has made a decided comeback, and is now thought to be the primary factor in the genesis of atherosclerotic plaques and thrombosis.

Sadly, interpretation of medical science is frequently influenced by the dark forces of industry, out to make a killing. And when there is money to be made, there will be big winners, but also big losers—including those killed. Witness the tobacco debacle. The risks of smoking have been known since the 1930s; the U.S. surgeon general report of 1964 squarely faced down the tobacco industry. That put the tobacco propaganda machine into overdrive to squelch the science and any scientists who stood in their way. My colleague at the University of California, San Francisco, Dr. Stanton Glantz was (and to this day still is) Public Enemy Number One of the tobacco industry. For twenty-five years he was a "prophet in the wilderness." Stan warned about Big Tobacco's tactics at every level: the political buy-offs, the marketing, the advertising to children, product placement in movies. He even uncovered blatant fabrication of data by the industry to exonerate their product. What did it get him? Twenty-five years of constant battles, both in the courtroom and in the court of public opinion. He was painted as a "false prophet, a zealot." But Stan had the courage of his convictions. More importantly, he had the data. Of course he was, and still is, right on target.

Indeed, who determines the difference between a prophet and a heretic? Whoever gets to write the history. It's only with our retrospectoscope that we seem to have twenty-twenty vision. Ask Galileo.

And so it is with Dr. John Yudkin. Let's set the stage. In 1955 President Eisenhower experienced a heart attack while in office. The issue of heart disease and its prevention was thrust into public consciousness. What component of diet caused heart disease? This was the seminal issue in public health, disputed in academic circles and the media throughout the 1960s and 1970s. Two factions sprang up. Dr. Yudkin was a University of London physiologist, nutritionist, and physician, and the primary exponent for the idea that sugar was the dietary factor promoting heart disease, and several others as well. First published in 1972, and updated with new science in 1986, *Pure, White, and Deadly* was, is, and remains, a prophecy. Yudkin foresaw the sugar glut that ultimately arrived with the advent of high-fructose corn syrup. He preached in the wilderness, and no one listened. In the other corner, Ancel Keys was a University of Minnesota epidemiologist who, in 1953, first espoused the argument that saturated fat was the primary cause of heart disease, culminating with his volume *Seven Countries: A Multivariate Analysis of Death and Coronary Heart Disease* (Harvard University Press, Cambridge, 1980). The debate grew beyond the academic; the rancor got up close and personal, with Keys declaring in 1971: "It is clear that Yudkin has no theoretical basis or experimental evidence to support his claim for a major influence of dietary sucrose in the etiology of [coronary heart disease]; his claim that men who have CHD are excessive sugar-eaters is nowhere confirmed but is disproved by many studies superior in methodology and/or magnitude to his own; and his 'evidence' from population statistics and time trends will not bear up under the most elementary critical examination" (Keys, A., *Atherosclerosis*, 14: 193–202, 1971).

Three scientific findings of the 1970s undid Yudkin's case and sealed his fate. Firstly, by studying the genetic disease *familial hypercholesterolemia* (victims experience heart attacks as early as eighteen years old), Michael Brown and Joseph Goldstein discovered low-density lipoproteins (LDL) and the LDL receptor (which won them the Nobel Prize), leading to the hypothesis that LDL was the bad actor in heart disease. Secondly, dietary studies showed that dietary fat raised LDL levels. Thirdly, large epidemiological studies showed that LDL levels correlated with heart disease in populations. Slam dunk, right? *It's the fat, stupid.*

The Pharisees of this nutritional holy war declared Keys the victor, Yudkin a heretic and a zealot, threw the now discredited Yudkin under the proverbial bus, and relegated his pivotal work to the dustbin of history, as this book went out of print and virtually disappeared from the scene. The propaganda of "low-fat" as the treatment for heart disease was perpetuated for the next thirty years. And the cluster of diseases (obesity, diabetes, hypertension, lipid problems, heart disease) collectively termed the "metabolic syndrome" increased in a parabolic fashion under the canopy of the sugar industry and their propaganda machine.

But good ideas die hard. Larger studies started to demonstrate that serum triglyceride levels correlated with heart disease, with sugar consumption being the primary driver. And there wasn't one type of LDL, there were two: large buoyant LDL, driven by dietary fat, but which was neutral in terms of heart disease; and small dense LDL, driven by dietary carbohydrate, and which oxidizes quickly, driving atherosclerotic plaque formation (hardening of the arteries). The Atkins diet was now being taken seriously. Carbohydrates started to assume center stage in promoting metabolic disease, with sugar consumption implicated as the most notorious carbohydrate.

I stumbled upon Dr. Yudkin quite by accident in 2008. I was in Adelaide, Australia, giving a talk at the Australasian Association of Clinical Biochemists on my research into the role of sugar in the pathogenesis of metabolic syndrome. Dr. Leslie Bennett said to me, "Surely you've read Yudkin," and I admitted I hadn't. When I got home, I looked for *Pure, White, and Deadly*, and couldn't find it in our UCSF library or in any bookstore in San Francisco. Eventually I got it by interlibrary loan. I opened the book, and it opened my eyes. I already knew from my own work that sugar at our current rate of consumption is a medical disaster. But to learn that Yudkin foresaw what a problem sugar was thirty-six years earlier, and at a much lower dose (i.e., before the advent of high-fructose corn syrup and the two-liter bottle) was a true revelation. Indeed, I was a Yudkin disciple and I hadn't even realized it.

Yudkin didn't have the voluminous data that exist today. He had correlation, but not causation. He didn't have mechanism. He didn't know that sugar caused insulin resistance by being turned into fat in the liver through the process of *de novo* lipogenesis, or that sugar induced protein damage through the *Maillard* or *browning* reaction. He didn't know that sugar was weakly addictive, although he surmised it. Despite that, *Pure, White, and Deadly* draws direct lines between sugar and dental caries, gout, autoimmune disease, heart disease, and cancer. Indeed, it shows that sugar consumption and mortality rates go hand in hand.

In the face of the current science and nutrition explosion, and the fall of the low-fat hypothesis, Penguin Books UK has chosen to reissue this "old" book, which is "new" again. We are now almost twenty-seven years removed from Dr. Yudkin's 1986 update. Surely, with all we've learned, this book must now be obsolete, isn't it? Not at all. First of all, true prophecies don't go out of style. That's like saying Darwin's *The Origin of Species* is irrelevant because Darwin didn't know what genes were. Secondly, it is a signpost on a journey of pilgrimage. It provides you with perspective on where you've come from, and where you're going. And lastly, Yudkin correctly fingered the sugar and food industries for what they were, and still are. Those who don't understand history are condemned to repeat it—especially in the face of persistent propaganda. And this book *is history.*

I'm proud to be a Yudkin disciple, to contribute to resurrecting his work and his reputation, and to assist in the advancement of his legacy and public health message. Every scientist stands on the shoulders of giants. For a man of relatively diminutive stature and build, Dr. John Yudkin was indeed a giant.

Introduction to the 1986 edition

A great deal has been written about sugar. There are dozens of books about the cultivation of the sugarcane and the sugar beet, including books that describe the shameful story of the slave trade between Europe, West Africa, and the Caribbean. There are dozens of books giving the technical details of sugar refining and the manufacture of sugar-containing food and drinks. But further accurate information about sugar as a food is not easy to come by. How many people eat more than average and how many eat less? Who are the small consumers and who are the big consumers and what are the smallest and largest amounts consumed? What would it do to our health if we took no sugar at all, or if we ate quite large amounts?

Part of this information can, with some trouble, be found in trade publications, but not all of it. You might think you could get it from the sugar industry itself; they undoubtedly have active information centers in many countries. We know what the average sugar consumption is in each country. But it is not possible to get the answer even to such simple questions as how much sugar is in the diets of people of different ages, or what is the range of the sugar content of the diet of 15-year-old British schoolchildren. It may be that the industry simply does not have this information, or it may be that they have it but do not wish it to be known. Especially, we would expect the sugar industry to be knowledgeable about levels of consumption when, in rejecting criticisms of the effects on health, they constantly refer to "moderate" consumption. Yet what the industry considers moderate must, on any reckoning, be quite a sizable quantity. One of the scientists who most strongly supports the sugar industry has written, "The usual range of sugar intake may therefore be between 10 and 30 per cent of total calories, with the average at 15 to 20 per cent." He goes on to say, "This rate of sugar intake may be considered moderate, and can probably be exceeded somewhat without over-stepping the balance of moderation."

Much more research has been done on the effects on health of the bread in our diet, or the eggs, or the breakfast cereals, or the meat, or the vegetables, than about the effects of sugar, even though sugar on average constitutes about 17 percent of our diet, a larger proportion than any of these other items. Yet in 1972, when *Pure, White, and Deadly* was first published, what little research there had been already showed that sugar in our diet might be involved in the production of several conditions, including not only tooth decay and overweight but also diabetes and heart disease.

Since that time research has produced further evidence that sugar is implicated in these conditions, and has also added to the list of diseases in which the sugar we eat may possibly, or even probably, be a factor. Many of the experiments from which these conclusions are derived have been carried out at the Nutrition Department of Queen Elizabeth College, University of London, some of them in collaboration with research workers in the Biochemistry Department. When our experiments have been repeated independently in other research institutes, the results have always been in line with our own. Those who disagree with what we say may therefore challenge the conclusions that we draw from the research, but they cannot legitimately disagree with the experimental results.

In this edition I have taken the opportunity to bring up to date and extend many of the statistics that I quoted earlier. I have also summarized the research that we and others have done during the last 14 or 15 years which has shown more of what happens in our bodies when we eat sugar.

I am often asked why we don't hear very much about the dangers of sugar, while we are constantly being told we have too much fat in our diet, and not enough fiber. I suggest that you will find at least part of the answer in the last chapter of this book.

John Yudkin

Pure, White,
and Deadly

1

What's so different about sugar?

Sugar is common enough in all our lives, and almost everyone believes that it is simply an attractive sweet—one of many carbohydrates in the diet of civilized countries. But sugar is really quite an extraordinary substance. It is unique in the plant that makes it, in the materials that chemists can produce from it, and in its use in foods at home and in industry. And recent research shows that it also has unique effects in the body, different from those of other carbohydrates. Since it now amounts to about one sixth of the total calories consumed in the wealthier countries, it is essential that more is known about what it does to people when it enters the body in food and drink.

Curiously enough, not only the layman but also the physician and the medical research worker have until recently assumed that there was no need to bother with any special study of sugar. Since man began to produce his food instead of hunting and gathering it, his diet has contained large amounts of carbohydrates of one sort or another (see p. 12). It did not seem to occur to anyone that it made any difference whether this carbohydrate consisted almost entirely of starch in wheat or rice or maize, or whether the starch was gradually becoming replaced by increasing amounts of sugar, as has been happening in the last 100 or 200 years.

Although some early research workers occasionally pointed out that eating sugar was not always the same as eating starch, no one paid much attention to this until 25 years or so ago. When I

1

wrote a book on weight reduction in 1958, I strongly recommended a diet low in carbohydrate, but I made very little distinction between the benefits of avoiding starch and avoiding sugar. Since that time, an enormous amount of new information has been accumulating, and more is being added constantly. Most of the new research has, quite properly, appeared in scientific and medical journals, but it seems now worthwhile to summarize it for nontechnical people. After all, it is not only scientists and physicians who eat, and if eating sugar really is dangerous, then everyone should be told about it.

The fact that so much about the effects of sugar is still being discovered is in itself an illustration of how unexpected it was to find so many differences in these effects from those of other common foods. You might have imagined that the realization that there were differences would have stimulated the sugar producers and refiners themselves to initiate studies into the properties of their product. Other industries which produce foods like meat or dairy products or fruits have spent a great deal of money over the years to carry out or support nutritional studies on their products, even though these foods form a smaller proportion of the Western diet than sugar now does. But the sugar people seem quite content to spend their money on advertising and public relations, making claims about quick energy and—as we shall see later—simply rejecting suggestions that sugar is really harmful to the heart or the teeth or the figure or to health in general.

I cannot claim that everything I say in this book will be accepted by every research worker. I hope however that I have made it clear which parts of the book refer to solid, observable scientific research and which parts are my own opinions and interpretations of these observations. Only time will show how right or how wrong I am in any one particular personal statement. But right at the outset I can make two key statements that no one can refute:

First, *there is no physiological requirement for sugar*; all human nutritional needs can be met in full without having to take a single spoon of white or brown or raw sugar, on its own or in any food or drink.

Secondly, *if only a small fraction of what is already known about the effects of sugar were to be revealed in relation to any other material used as a food additive, that material would promptly be banned.*

Take the case of cyclamates. Some countries now do not permit this sugar substitute to be used, and the prohibition is based on experiments in which rats were fed for an enormously long time on huge amounts of cyclamate—the equivalent of a man consuming 10 to 12 pounds of sugar every day for 40 or 50 years. Later in these pages you can read what can happen to rats fed sugar in amounts hardly—if at all—different from those consumed by very many people. I will not anticipate the details that you will find, but the very many effects include enlarged and fatty livers, enlarged kidneys, and a shortening of life span.

Think of all this the next time you read of an experiment that suggests that another sugar substitute may be harmful, as happened when aspartame was introduced. Note the blaze of publicity encouraged by the busy men and women who run such organizations as Sugar Information Incorporated or the Sugar Bureau. Then think of what is already known that sugar *can* do, as distinct from what the substitute might *possibly* do if taken in enormously unrealistic amounts for a long enough time.

My own view is that it is perfectly safe to use these sweeteners whenever you wish to, although (for what I consider quite inadequate reasons) you cannot find cyclamate in some countries. But although they are quite safe, some people think it a good idea not to use sweeteners. They prefer to get into the habit of having less sweetness in their foods and drinks, by avoiding those foods that *must* be made with sugar.

Many people have criticized what I have previously written; they say that the experiments that we and others have carried out have used absurdly high amounts of sugar to produce the effects we describe. One such person is the American physiologist Dr. Ancel Keys, the most important and certainly the most dogmatic research worker who expounds the view that coronary disease comes from dietary fat and that sugar has nothing whatever to do with it. He has written that "the level of sugar in the experimental diets are of the order of three or more times that in any natural diet." This is quite untrue, as we shall see, but it comes about because very few people have bothered to find out how much sugar people do, in fact, consume.

You hear stories that the Turks take a very great deal of sugar, as you can see from the amounts they put into their coffee. But the Turks even now only take about one half of the amount consumed in Britain and the United States, and 20 years ago the Turks took less than one quarter. Apart from these sorts of questions, you can also go wrong when you look at official statistics without reading the small print. There have been regular annual reports of the British diet for the last 40 years, and the figures given for sugar now amount to an average of about 32 pounds a year. But if you look carefully, you will see that the statistics do not include snacks or food eaten away from home, and the real average turns out to be more than three times as much, about 100 pounds of sugar a year. If you now take into account that this is an average, and that many people take much more sugar than the average, you will find that the quantities used in experiments with human beings and animals are by no means extraordinary or absurd.

And what about Dr. Keys's reference to the sugar content "in any natural diet"? What *is* a natural diet? Is it "natural" for Westerners today to eat 20 times as much sugar, or more, compared with what our ancestors ate only two or three hundred years ago, and vastly more than our earlier ancestors had ever eaten? Nowadays we hear so often the words "natural" and "moderate"; we really must be on our guard not to be misled into believing that they have any real meaning, or even worse that they provide evidence that something to which these words are applied is intrinsically wholesome, good, and desirable.

I hope that when you have read this book I shall have convinced you that sugar is really dangerous. At the very least, I hope I shall have persuaded you that it *might* be dangerous. Now add to this the fact—the indubitable fact—that neither you nor your children need to take any sugar at all, or foods or drinks made with it, in order to enjoy a completely healthy and highly nutritious diet. If as a result you now give up all or most of your sugar eating—and I shall show you later that this is not too difficult—I shall not have wasted my time in writing this book, and more importantly you will not have wasted your time in reading it.

2

I eat it because I like it

One of the most spectacular current "growth industries" is that concerned with the production and distribution of health foods. In Britain and the United States almost every neighborhood has its special store where you can, it seems, ensure eternal youth by buying handwoven honey, free-range carrots, and stone-ground eggs.

There is no doubt that people today are very worried about their food. But different people are worried about different things, and most of them are worried about the wrong things. I can assure you that it really does not matter to your health whether your chicken is produced by the broiler system, or whether you eat potatoes grown with chemical fertilizers. But it *does* matter that your diet is now very likely to be different from that which has been evolved over millions of years as the diet most suitable for you as a member of the species *Homo sapiens*.

Please don't take these sentences to imply that I have discovered the secrets of the ideal diet. Because I have written rather teasingly about "natural foods," I do not mean to imply that everything you see in the health food store is nonsense and that everything I shall be telling you is an absolute certainty. It is true though that every person tends to believe that a knowledge of nutrition is somehow instinctive and that careful thought and introspection will provide as good an answer to nutritional questions as do the studies and research of the professional nutritionist.

It is silly to insist, in spite of all the detailed evidence to the contrary, that there are any differences in the nutritional value of potatoes produced on land fertilized by chemical fertilizers or by compost. On the other hand, it is equally silly of some scientists to imagine that we know all there is to know about human nutrition. There is, for example, no justification for the statement I heard at a scientific meeting, where a food chemist said that scientists don't have to concern themselves too much about producing enough high-protein foods; human beings will soon be able to feed themselves entirely with synthetic protein and other nutrients. And this at a time when new facts are discovered almost daily about such supposedly well-understood phenomena as obesity, or about the effects of different dietary carbohydrates. The safest position is somewhere between arrogance based on unrecognized ignorance, and arrogance based on unwarranted certainty.

But how do we find this position? What sorts of principles do we adopt in order to decide whether this or that food is "good for you"? What indeed should the ideal diet be?

I am going to devote the rest of this chapter to trying to answer these questions, slowly and carefully, because I believe that an understanding of the biology of the diet provides the clues to what the Western diet should be; what is wrong with it today; and why it has gone wrong.

We begin by reminding ourselves that all animals require two sorts of materials for their growth and survival. One is material that can be burned (oxidized) to yield the energy needed for the processes of living—growth and movement and breathing, and all the other activities that distinguish a living animal from a dead one. These materials for energy production are mainly carbohydrates and fats, although protein can also be used in this way. The second sort of material consists of those thousands of different compounds that go to make up the very complex chemical composition of the cells of the different tissues that, organized together, constitute the whole living animal. The vast majority of these compounds can be made by the body itself, from a very much smaller number of raw materials. But these are all materials that must, each one of them, be supplied to the body. Without them, a young organism cannot grow, and an adult organism will gradually waste away because it is unable to make good the general wear and tear of its cells and tissues.

So we can say at this point that the body has to be given materials both to supply energy and to provide the raw materials for growth and repair. The source of these essential materials is our food and drink. These have to supply about 50 different items. They fall into several classes—the carbohydrates, the fats, the proteins, the vitamins, the mineral elements—and of course water.

As far as we know, every single species of animal needs the same components for life and sustenance. And almost every single species has to get all of these out of food. The exceptions are interesting, and include ruminants like cows which can get many vitamins from microbes living in their complicated stomachs. But in general, as I said, most animals have to get all of their vitamins, protein, and so on from their food, and these nutrients are needed in roughly the same proportions by all animal species.

You could therefore argue that all species of animals should eat the same foods. But in fact it is well known that different species eat very different diets indeed. Some, like the lion and the tiger, are largely carnivorous—meat-eating. Others, like rabbits and giraffes and deer, are largely herbivorous—plant-eating or vegetarian. Others again, like ourselves and rats and pigs, eat diets that come from both animal and plant sources; these animals are omnivorous. By contrast some animals eat only a very limited range of foods; the giraffe eats little except leaves from acacia trees. The koala bear eats little except eucalyptus leaves, and then only from a few of the 400 or so existing species.

So there is an apparent contradiction. First, all species of animals require the same in the way of nutrients, which—with a few exceptions—they must get from their food. But secondly, different species of animals get these same nutrients from very different sorts of diet. Great biological advantages flow from this, because it prevents the various species competing with each other for the same foods. Each species establishes its own "ecological niche" in regard to its food supply. Its anatomy and physiology are well adapted to find, acquire, eat, chew, and digest the foods that it chooses.

But the fact remains that one species will often not even attempt to eat foods that are highly sought after by another species. So what makes one animal choose one sort of diet, and a different animal choose a completely different sort? Clearly, it cannot be that they are choosing these different foods for the nutrients they contain, since their nutrient needs are so similar. It must therefore be some other properties of foods that make one range of foods look especially attractive to one species, and another range especially attractive to another. These qualities are shape and size, color and smell, taste and texture—features that I'd like to lump together, perhaps too loosely, under the heading of palatability.

Foods thus possess two different properties—palatability and nutritional value. The palatability of foods, and so the foods chosen to make up the total diet, varies from species to species; however, the nutritional needs that have to be satisfied by these various species are virtually the same for all species. Thus, animals choose diets that they find palatable, but, whatever these diets are, they must supply all their nutritional needs. If they did not, the animals would perish.

So we can say that when an animal eats what it wants, it gets what it needs; or, in terms I have just been using, for each sort of animal palatability is a guide to nutritional value. Everyone instinctively feels that this is correct; if you like some food very much it is taken to indicate—to prove, almost—that you need this food.

Eating habits are formed in childhood, and children like sweet foods. Does it follow that sugar must be good for them? Not at all, although I am sure that most people have heard this sort of argument. One also hears phrases like the one in the old music hall song, "A little of what you fancy does you good." And so long as human beings did not manufacture foods, this argument was perfectly sound.

The origin of the human diet

I shall come back later to the question of when it is true that what you want is what you need, and when it is not true. Let me now pick up the story of palatability and nutritional value, and see how it applies to our own species.

Science is gradually learning quite a lot about our origins, and although there are still a lot of uncertainties about the early human diet, one can now make some pretty good guesses.

It is generally agreed that our earliest ancestors, the squirrel-like primates of some 70 million years ago, were vegetarian. They continued as vegetarians up to about 20 million years ago, for they had no difficulty in surviving on fruits, nuts, berries, and leaves. But then the rainfall began to decrease and the earth entered a 12-million-year period of drought. The forests shrank and their place was taken by ever-increasing areas of open savannah. It was during this time that *Australopithecus africanus* emerged. (*Australopithecus* means "southern ape.")

In order to survive, *africanus* had to forsake the vegetarian and fruitarian existence of the related hominid *Australopithecus robustus*, and change to a scavenging and hunting existence that was largely carnivorous. The molar teeth of *africanus* had the shape and thin enamel of a carnivore. The jaw muscles were small and did not need the crested cranium of *robustus* for their attachment. The canines were also small, for *africanus* killed neither with fangs nor with claws or horns, but with weapons, having adopted a completely erect posture, which freed the arms and hands from the need to be used for locomotion. *Africanus*'s earliest weapons were bones; only later did stones begin to be used, and still later the axe.

Thus it appears that for at least two million years our ancestors were largely meat-eating. From that time, they continued to be scavengers and hunters, seeking their favorite food of meat and offal.

They had one advantage over the more strictly carnivorous species, in that they could and did eat vegetable foods too. Along with meat, their diets contained the nuts, berries, leaves, and roots that had fed their forebears. This omnivorous potential gave them the ability to survive when their prey eluded them or was scarce.

In nutritional terms, the diet of prehistoric human beings and their ancestors during perhaps two million years or more was rich in protein, moderately rich in fat, and usually poor in carbohydrate. If we assume that our present universal taste preferences for the sweet and savory are a continuation of preferences acquired long ago, then it is likely that, except in times of hunger, the small amounts of dietary carbohydrates will have come mostly from fruits, as opposed to the less palatable leaves and roots.

The two food revolutions

Until very recently in evolutionary terms, all animals, including human beings, depended for their food supplies on hunting or scavenging other animals, or on the consumption of wild vegetation. It was less than 10,000 years ago—compared with the two million years or more of carnivorous ancestry—that we became, uniquely, food producers. Agricultural food production seems to have originated independently at three different times in three different parts of the world, from which it then spread. The first was around 10,000 years ago in the Fertile Crescent, in what is now Israel, Jordan, Syria, Turkey, and Iran, with the cultivation of wheat, barley, lentils, and peas, and the domestication of cattle, sheep, and goats. About 7,000 years ago agriculture began in China, producing rice, soybeans, yams, and pigs. The area that came last to agriculture was Central America, where the chief crops were maize and beans, and where llamas and guinea pigs were raised.

In most instances, then, food production began with the cultivation of cereals. This derived from the discovery that some of the wild grasses whose seeds were occasionally eaten could yield many times that amount of edible seeds if they were deliberately planted. The domestication of these grasses produced the cereals that are now the staple food of a large part of present-day humanity and it was followed or accompanied by the domestication of root crops, and of wild animals that were used both for food and as animals of burden.

The results of the discovery of agriculture—the Neolithic revolution—were many and far-reaching. Human beings ceased being nomads and began to live in settled socially organized communities. This landmark of progress became the basis for all that we know of civilization, with its arts, its inventions, and its discoveries.

Compared with hunting and foraging, agriculture usually yielded more food; it also allowed the cultivation of areas where existing resources of food would have been inadequate. Thus the human population grew, because fewer died of food shortage and because people spread into increasing areas of the earth's surface. But in due course the limits of food production again became the limits to the numbers that could be fed. The inevitable pressure of population on food supplies tended to produce and stabilize a type of diet quite different from that of our hunting ancestors. It was— and still is—much easier to produce vegetable foods than animal foods; for a given area of land, some ten times as many calories can be produced in the form of cereals or root crops than in the form of meat, eggs, or milk.

The effect of the Neolithic revolution was thus to alter the components of the diet so that it was now rich in carbohydrate and poor both in protein and in fat. The carbohydrate was overwhelmingly starch, with sugars supplied only to a small extent as before by wild fruits and vegetables. It is likely that deficiency of protein and of many of the vitamins began to affect large sections of the human species only after they became food producers.

Human beings, like all animals, constantly face recurring periods of food shortage. Although the Neolithic revolution increased total food supplies and radically changed the composition of our diet, hunger and famine did not vanish. For most of the time, wind, drought, flood, and our own exploitation of the land have combined to limit food production to levels lower than those necessary to feed all our offspring. It is only in the last few decades that a sizable proportion of people—though still only a minority—have been born into a situation where it is likely that they will never know real hunger throughout their lives.

The reasons for this second revolutionary change are the cumulative effects of science and technology. I need only list a few of these to show the extent of this revolution and its effect upon the availability of food to mankind: genetics and the breeding of improved varieties of plants and animals for food; engineering and its effect on drainage and irrigation; the discovery of synthetic fertilizers, weed killers, and pesticides; the internal combustion engine and its effect upon transport by sea, land, and air; modern methods of food preservation by canning, dehydration, deep freezing. I could cite many more examples of changes that have given humanity the possibility of producing and preserving much more food than has ever been available to any other species.

As a result, in the affluent countries a large proportion of the populations has a very wide choice of foods, irrespective of season or geography. The effect has been that these people are able more and more to choose foods that please their palates, and not simply foods that fill their stomachs. The first and most obvious result has been an increase in the consumption of more palatable foods, such as meat and fruit. And because of the basic association between palatability and nutrition, there has come a simultaneous improvement in the nutritional standards in these groups, just as there has always been a better level of nutrition in the much smaller section that comprises the wealthy members of any population.

The advances in agricultural techniques and general technology have had an effect not only on the yield of food and the availability of food. They have also had a tremendous effect on the way foods can deliberately be changed by extractions and additions, so that quite new foods can be made that do not exist in anything like these forms in nature. Some of these manufactured foods have been in existence for quite a long time—bread, for example, and tortillas and chapatis and cakes and biscuits. But most of them have been produced, or vastly improved, only in the past century or two or in recent decades. I am thinking now of ice cream and soft drinks, an enormous range of chocolate and confectionery, and new sorts of snacks in the form of sweet and savory biscuits. And there is now a new range of "meat" products made from textured vegetable or microbial protein.

We can do all these things largely because nutritional value and palatability are two different qualities. As I pointed out, although we can use as food almost any sort of animal or vegetable material, our preferences are for the particular palatability qualities of meat and of fruit, which together can supply all the nutrients we require. We are only just beginning to emulate the taste and texture of meat; and people will be eating and relishing significant quantities of the new vegetable or microbial protein foods only when the food manufacturer imparts to them qualities that make them much more attractive than he has been able to do up to now. But for some time industry has been able to isolate an essence of sweetness, which has the property of imparting a very desirable palatability to a wide range of foods and drinks. People do not demand a particular flavor and texture to go with sweetness, although they seem to demand only a very limited range of flavors and textures to go with savory foods.

The human avidity for sweetness could for vast periods of time be satisfied almost exclusively by the eating of fruit; rarely, and in very small quantities, our ancestors might be lucky enough to find some honey produced by wild bees. But some time after the Neolithic revolution, perhaps only 2,500 years ago, people found that they could produce a crude sort of sugar by extracting and drying the sap of the sugarcane. This first began to be cultivated probably in India, and its cultivation slowly spread to China, Arabia, the Mediterranean, and later to South and West Africa, the Canary Islands, Brazil, and the Caribbean.

In spite of this increasing area of cultivation, the cost of the sugar, crude as it was, was extremely high, so that by the middle of the sixteenth century it was said to be equivalent to the present cost of caviar. Compared with the price of foods such as butter or eggs, it has been calculated that the price of sugar has fallen to about a two-hundredth of its price in the fifteenth century. Even as late as the eighteenth century, sugar was a luxury, and until a hundred years or so ago domestic sugar boxes were often provided with lock and key.

It was chiefly the development of the sugar plantations in the Caribbean, based on the slave trade, that set the pattern of the sugar industry in the form known today. The demand for sugar was so great, and its production so lucrative, that tremendous improvements began to be made from about the middle of the eighteenth century in the production of high-yielding sugarcane (and later the sugar beet); in the efficiency of the extraction of the sugar and the making of raw sugar; and finally in the process of refining the sugar. Thus, the price fell constantly, the demand grew, and consumption rose to exceedingly high levels.

Legislators in many countries have often taxed sugar to provide revenue, just as they have often taxed tobacco and alcohol. And sugar also resembles alcohol and tobacco in that it is a material for which people rapidly develop a craving, and for which there is nevertheless no physiological need.

I am saying, then, that human beings have a natural liking for sweet things; that primitive people could satisfy this desire by eating fruit or honey; and that in eating fruit because they liked it, they obtained necessary nutrients such as vitamin C. But now we can satisfy the desire for sweetness by consuming foods or drinks that provide little or no nutritional value except calories. It is possible today to get an orange drink that is more attractive in color than true orange juice, is sweeter in taste, has a more aromatic flavor, is cheaper to buy—and can be guaranteed to contain no vitamin C whatever.

Since people chiefly seek palatability in foods and drinks, the sale of these drinks increases all the time. One day it will no doubt be possible to manufacture from some non-digestible polymer a hamburger that looks more attractive than a real meat hamburger, and smells and sizzles better on the barbecue, at only half the price. It will be entirely "pure" in that it will contain neither protein nor vitamins nor minerals. And who will say that we shall not buy this super, space-age, new food just because it has no nutritional value. We shall buy it because we like it, and only because we like it.

Most people still believe that foods that are palatable must have a high nutritional value; many also believe what is equally untrue: that foods with little flavor have no nutritional value. I am certain that it is the dissociation of palatability and nutritional value that is the major cause of the "malnutrition of affluence." For this reason, let me give you one or two more examples of how one can no longer expect the two qualities to be found together.

First, you may remember beef tea, which even in this century was commonly given by doctors to their convalescent patients as a "restorative." And to this day many mothers believe that a tasty clear soup is nourishing for their children. Yet here is high palatability with virtually no nutritional value. Second, the economics of chicken farming has produced a broiler chicken which, because it is slaughtered young, and because of the speed with which it is eviscerated, has less flavor than a free-range chicken. Yet its nutritional value is no different, even though its lower palatability is often referred to as indicating a lower nutritional value.

Some time ago I read a short story, the title and author of which I have unfortunately forgotten. A brilliant chemist became tired of his mistress and decided to get rid of her by using his professional skill. He devoted himself to developing a new and exquisite flavor, which he then incorporated into chocolates, sending box after box to his mistress. Finding these quite irresistible, she consumed them in inordinate quantities until she died of overeating. The chemist knew that her craving would alone suffice to kill her.

One more example of the strong power of palatability is the story of the snake that ordinarily will eat only toads. It will not, for example, eat pieces of meat such as beef. But you can make it do so by rubbing the beef onto the skin of the toad and so presumably making the beef taste of toad.

One argument used by the health food people to demonstrate the poor nutritional value of modern processed foods is to claim that they have little flavor. Their own products, they say, *must* be nutritionally superior because they taste better. Much of what I have to say in this book is based on the proposition that satisfying our palates is no longer a guarantee that we are satisfying our nutritional needs.

3

Sugar and other carbohydrates

The different sorts of sugar, including sucrose and glucose, all belong to a group of substances known as carbohydrates. Since we shall be talking of these substances from time to time, let us look at the whole group for a moment.

The carbohydrates in our diet can be divided into those that the body can digest and absorb from the gut and those that cannot be absorbed; they are sometimes referred to as "digestible and indigestible," or "available and unavailable." The unavailable carbohydrates, which pass through the body virtually unchanged, make up the greater part of what is now known as fiber and used to be called roughage. This mostly consists of cellulose, the chief constituent of cotton and of paper.

The available or digestible carbohydrate of the diet consists almost entirely of sugars and starch. They are all made up of units called monosaccharides. Chemists apply the word "sugar" to any one of a particular group of substances that have similar properties but are not identical. Some of the better known sugars are glucose, fructose, maltose, lactose, and sucrose; these are either monosaccharides or disaccharides.

The best-known monosaccharides—sugars made up of single units—are glucose, fructose, and galactose. Glucose is the first product of photosynthesis in plants, and is the main source of energy for both plants and animals. Fructose, together with some glucose and sucrose, is found in fruits. Galactose exists only in the animal kingdom, as part of milk sugar—lactose.

Glucose is a sugar found, usually with other sugars, in some fruits and vegetables. It is very important to biochemists, biologists, and nutritionists because it is a key material in the metabolism of all plants and animals. Many of our principal foods are sooner or later converted in the body into glucose, and it is one of the most important substances that is metabolized (oxidized or burned) in the tissues to supply energy for everyday activities.

There is always glucose in the bloodstream, and this is usually called "blood sugar." In healthy people, a complicated interaction of a number of hormones contrives to keep the level of the blood sugar fairly constant. If you eat ordinary sugar or starch, or one of several other substances, glucose will be released during digestion and this will be absorbed from the alimentary canal into the blood. The level of blood glucose therefore rises. Immediately, however, there is an outflow of hormones, especially insulin from the pancreas, into the bloodstream; the effect of this is to lower the level of glucose toward its normal level. This works chiefly by converting it into a polysaccharide (made of many monosaccharide units) called glycogen and tucking this away in the muscles and liver, where it can be called upon again to release glucose if the level in the blood falls.

Sucrose, the chemical name for the subject of this book, is one of three common disaccharides. It is made up of one unit of glucose joined to one unit of fructose. When digested, a mixture of equal amounts of glucose and fructose, called "invert sugar," is produced. There is reason to believe that it is the fructose part of sucrose that is responsible for many of the undesirable effects of sucrose in the body.

There are two other disaccharides to be found in human diets. One is maltose, made up of two units of glucose joined together. This is produced during the digestion of starch, for example when grain like barley begins to germinate, or when starch is in the mouth being chewed, or when it reaches the intestine. It is later digested to glucose. The third disaccharide is lactose, produced by the joining together of two monosaccharides, glucose and galactose. It occurs only in milk, or in foods such as yogurt that are made from milk and that include the water part. Large quantities of lactose cause diarrhea, and even relatively small quantities have this effect on people with lactose intolerance. However, such people are not affected by small amounts of milk—say, up to a pint a day taken at intervals. They also tolerate cheese, since most of the lactose remains in whey when cheese is made.

Starch, which occurs as a store of energy in plants, is made up of many glucose units joined together and is therefore called a polysaccharide. It is easily digested either by enzymes in the body or enzymes extracted from molds, or by heating in solution with acid. The effect is to break down the starch into smaller and smaller pieces. The early stages result in the production of dextrins. Later maltose is produced, and finally glucose. Glycogen, as we saw, is another polysaccharide, found in the liver and muscles of animals. Like starch it is a store of energy, but unlike starch it is present in relatively small quantities: the total amount of glycogen in an adult human body is no more than 350 grams. Cellulose is also a polysaccharide, but is not digestible.

4

Where sugar comes from

The sugar with which this book is concerned is what most people simply call "sugar." It is sometimes called "cane sugar," although in fact about one third of the sugar consumed comes from beet. Chemists call it "sucrose." In this chapter we discuss where it comes from and how it is prepared; we shall then look at the effects it can have in the body.

Nearly 99 percent of the sugar we consume is made from sugarcane or sugar beet. The other 1 percent comes from such sources as the maple in New England and Canada, the palm in India, and millet in the southern United States; quite tiny quantities are also sometimes made from grapes, carob beans, or dates.

In spite of popular belief there is no difference in taste or any other ordinarily recognizable property in refined sugar isolated from the cane or the beet. Each contains more than 99.9 percent pure sucrose. Only the most sensitive analytical techniques can detect differences due to the presence of minute quantities of substances characteristic of one or the other.

Cane sugar

Like the cereals, the sugarcane (*Saccharum officinarum*) belongs to the grass family. Its original home was apparently somewhere in Asia, possibly India. It has been cultivated for 2,500 years and its wild ancestor is no longer known. It now grows mostly in plantations in many parts of the world. Cultivation probably began in China and India before 500 BC; in 325 BC the soldiers of Alexander the Great in India spoke of "honey not from bees." Cultivation spread westward, reaching Persia (by AD 500), Egypt (640), Sicily and Cyprus (700), Spain (755), and later Madeira, the Canary Islands, North America, Mexico and—by the beginning of the sixteenth century—the Caribbean. The cane requires a warm climate, rainfall of at least 60 inches a year or adequate irrigation, and plenty of fertilizer. The countries producing the greatest quantities of cane sugar are shown in the table.

World sugar production (1982)
(million tonnes)

Total 101	Cane Sugar 64		Beet sugar 37	
	India	9.1	EEC	13.4
	Brazil	8.9	USSR	7.0
	Cuba	8.0	USA	2.9
	Australia	3.7	Poland	1.9
	China	3.0	Turkey	1.6
	Thailand	3.0	Spain	1.1
	Mexico	2.7		
	Philippines	2.7		
	USA	2.5		
	South Africa	2.4		

The history of sugarcane cultivation in the Caribbean can hardly be a source of pride to humanity. The Europeans—from Portugal, Spain, Holland, and Britain—who first took the sugarcane to the West Indies rapidly overcame the indigenous population of Caribs and then proceeded virtually to exterminate them. They solved the problem of providing the labor force needed in the sugar plantations by bringing slaves from Africa. Thus was established the infamous "triangular trade." Rifles, cloth, and other goods were shipped to the west coast of Africa where they were given to the African chiefs in exchange for slaves captured in the interior. These were herded into the holds of ships and taken to the Caribbean Islands such as Jamaica and St. Kitts. Those that survived the horrendous conditions of the journey—often less than half—were then used on the plantations. The raw sugar from the islands was shipped back to Europe—especially to England—for refining, thus completing the triangle. The appalling conditions in the ships and the plantations caused so many deaths that the supply of slaves had to be constantly replenished by fresh imports from West Africa.

In the earliest days cane sugar was simply the dried juice that had been pressed from the cane; a similar product known as "gur" or "jaggery" is still made in India. Most sugar used nowadays, however, is refined white sugar. Sugar from the cane is usually produced in two stages—the extraction of raw sugar, and then the refining of this to white sugar. The sugarcane is cut by hand or, increasingly, by machine; the tops and leaves are removed and the canes brought rapidly to the factory. There they are cut, crushed, shredded, and passed through roller mills which press out about two thirds of the juice. The crushed fiber, known as "bagasse," is sprayed with a little water and passed through another set of roller mills. The quantity of dried bagasse produced is more than enough to provide for the energy needs of the factory; the surplus is sold to local generating stations or for paper manufacture, or mixed with molasses for animal feed.

The juice pressed out of the cane is at this stage a turbid grayish liquid containing some 97.5 percent of the sugar originally present. About 16 percent of the juice consists of dissolved or suspended solid matter, of which 85 to 90 percent is sucrose. The juice is now heated to boiling point and lime added. This produces a copious precipitate, which rapidly settles as "bottoms," leaving clarified juice above. The "bottoms" are spread over the fields as "mud," acting as a fertilizer. The clarified juices are evaporated, first in open vessels and then in vacuum pans. Eventually the sugar begins to crystallize, becoming "massecuite," a mixture of sugar crystals and syrup. These are separated by spinning in a centrifuge at up to 1,200 revolutions per minute.

The result is two products—raw sugar and cane molasses, or syrup. The molasses are boiled twice more, and the process of crystallization and centrifugation is repeated. After the third boiling there is usually not enough sugar left in the molasses to make it worth trying to extract more as crystals. The final molasses, with whatever sugar it does contain, is used in a variety of ways—for example, to make rum or yeast or cattle food.

The raw sugars produced by the three boilings are progressively darker in color: the lightest is called demerara, even though most of it no longer comes from the part of Guyana bearing that name; the second crop of crystals is called light muscovado and the third crop dark muscovado.

The next step is for the raw sugars, either separately or mixed together, to be sent to the consuming countries, where they are processed into refined white sugar.

The raw sugars are washed and then dissolved ("melted") in water, and the solution is decolorized by passing it through columns of charcoal. It is then put into vacuum evaporators and boiled until the concentration of sugar becomes high enough for crystallization to take place. This is started when a small quantity of crystals is thrown into the concentrated syrup, the precise timing determining the size of the new crystals. The contents of the evaporator, a mixture of crystalline sugar and syrup, are now transferred into large centrifuges, each with an inner perforated basket in an outer cylinder. Rotation of the basket results in the syrup being forced through the perforations while the crystals remain behind. If sugar cubes are required, the wet mass is poured into a shallow flat tray which is covered with a lid and passed slowly through a heated chamber. The resulting thin slab of sugar is chopped into cubes by a sort of multiple guillotine.

Some cane juice, instead of being evaporated to make raw sugar, may be processed in the plantation factory, producing what is known as "plantation white." This may be done by "sulfitation," in which sulfur dioxide is passed into the limed juice so that calcium sulfite is formed.

Alternatively, calcium phosphate may be produced in the juice, or both phosphate and sulfite. Rarely, the cane sugar juice is treated by the carbonation process which is routinely used for beet sugar juice. Whichever method is used, the virtually clear, colorless juice is filtered off and dried in evaporators to give plantation white sugar.

Beet sugar

The sugar beet, *Beta vulgaris* (subspecies *circla*), grows as a white root, and is related to the common red beetroot and to the mangold. It grows well in temperate climates, requiring a deep, limy loam that is well drained. The discovery that sugar beet might be a source of sucrose was made by the German chemist Marggraf in 1747. It was, however, not until the Napoleonic Wars that another German, Achard, working in France, demonstrated that it could be refined on a commercial scale. Its main advantage was that, unlike sugarcane, beet could be grown in temperate climates, and France began producing beet sugar in 1811 to avoid the effects of the Allied blockade which was preventing the import of cane sugar.

Since the molasses from sugar beet is so bitter as to be unacceptable to the human palate, no attempt is made to extract raw sugar from the beet; processing is a single operation leading directly to the production of refined sugar. First the washed beets are sliced into strips, or "cosettes." The extraction of the juice is carried out by diffusion, in a series of a dozen or so cells. The sliced beets pass along from cell to cell in one direction, while water enters at the other end and passes from cell to cell in the opposite direction. Thus, at one end, fresh beet slices enter and sugar-rich juice is withdrawn, and at the other end fresh water is admitted and the exhausted beet slices discharged. The juice then goes through the process of refining in the same way as for cane sugar. Brown sugar can be produced by mixing some cane molasses or caramel with the refined white beet sugar, as is sometimes done with refined cane sugar.

About half of the white sugar consumed in the UK comes from the sugar beet, the other half from the cane.

5

Is brown sugar better than white sugar?

By far the greater part of sugar from the cane ends up as refined white sugar; a small quantity is sold as brown sugar. But not all the brown sugar available to the consumer is this unrefined raw cane sugar. Some, as we have seen, is manufactured from white refined sugar—either from cane or beet—by the addition of molasses or caramel. Unfortunately, it is legally permissible to describe as "demerara" the light brown sugar made in this fashion, which bears a superficial resemblance to the raw sugar produced in the first boiling.

The characteristics of the unrefined raw sugars depend on several factors. First, there is an increasing proportion of molasses trapped within the sugar crystals as the syrup passes from the first to the third crystallization, producing first demerara sugar, then light muscovado and dark muscovado. Thus, each successive sugar is of a deeper brown color (accentuated by the greater degree of caramelization caused by repeated boiling and crystallization) and has a stronger flavor of caramel and molasses (known as treacle in Britain).

But other factors are also involved. Strains of sugarcane produce juices containing differing amounts of substances with various undesirable qualities, some of which will adhere to the sugar during crystallization. By choosing an appropriate strain of cane and taking care to keep extraneous materials out of it when it is harvested and cut, the raw sugar produced can be made to consist of clean, evenly sized, bright crystals with an attractive brown color and a pleasant taste and aroma. Without these standards of diligence and care, the same general process can yield a dirty product containing easily observable nonsugar particles mixed with uneven particles of dull brown sugar, the whole having an unattractive aroma. This is especially noticeable in the crystallization of the dark muscovado, but may be detected too in demerara. This does not matter if the raw sugar is produced as an intermediate stage on the way to the refinery. However, some of this dirty raw sugar, not really fit for consumption, is put on the market side by side with the clean raw sugar intended from the outset to be consumed in the unrefined state. You can see the difference in quality if you closely examine just a teaspoonful, placed on a white saucer and shaken into a thin layer.

Careful inspection will also show the difference between these unrefined sugars and the brown sugars made by adding molasses to white sugar. With the latter, you will notice that the color is only on the surface, and a quick rinse with a little water will reveal the white crystals of sucrose. In the UK, however, this kind of testing should be unnecessary, since labeling makes it easy to distinguish these two kinds of brown sugar. The raw sugars are labeled as "unrefined" or "raw," and the country of origin is given. The colored white sugars have to be labeled so as to indicate their ingredients. The wording will be something like "Ingredients: cane sugar, molasses." These sugars are also likely to be given some such description as "light brown" or "dark brown" or "London demerara" or "golden granulated."

There was a time when brown sugar, like brown bread, was considered to be less pure and less desirable; it was also less costly then. As a result it was the wealthier people who ate white bread and white sugar, while it was an aspiration of the less wealthy to be able to do the same. But from time to time a minority took the view that, far from the brown color indicating a degree of impurity, it indicated that the food was better because it had not been deprived of some important nutritious components.

Unlike brown bread, however, which is almost always bread

made from flour produced from whole wheat or lightly milled wheat, a great deal of the available brown sugar is made, as we saw, by the addition of caramel or molasses as a coating to crystals of refined white cane or beet sugar. Many of those who buy brown sugar do so in the belief that they are buying raw sugar; this does not matter much if the brown sugar is bought because of its taste. The situation is altered, however, if it is bought in the belief that it retains some nutrients that are removed when raw sugar is refined.

The conventional view of the nutritionist used to be that neither colored white sugar nor raw sugar contains anything that gives it a significantly higher nutritional value than that of refined sugar; I too held this view when I wrote the first edition of this book. Since that time, however, my colleagues and I have carried out a series of experiments that showed that at least some raw sugars may contribute to the nutritional value of a diet.

We decided to do these experiments because of the publication in 1981 of a series of reports describing research carried out in several laboratories in the USSR. These compared the effects in rats and mice of feeding diets containing either white (refined) sugar or brown (unrefined muscovado) sugar. The animals fed the brown sugar were reported as showing more rapid growth, prolonged life, less increase in the concentration of cholesterol in the blood, larger litters, and a better metabolic picture, especially in relation to carbohydrate metabolism. The Soviet workers claimed that these beneficial properties of the brown sugar resided in a number of complex organic substances to which they gave the name "biologically active substances" (BAS).

Their results were sufficiently striking for us to examine these claims in our own laboratory. We made up our usual sort of laboratory diet, which contained protein, fat, vitamins, and mineral salts, together with either refined sugar or brown muscovado sugar or pure starch. We fed our rats from the age of three weeks with one or other of these diets. Our results were very different from those reported by the Soviet workers. We could not confirm their claims; the different sugar diets produced the same growth rate, size of litters, and carbohydrate metabolism. The only differences were the usual ones we had discovered between rats fed sugar and rats fed starch.

After about two years of experiment we were about to discontinue our research when we decided to carry out one last investigation. We thought it would be interesting to see what the effect was, not simply on the rats themselves but on their pups. We therefore allowed the pups to stay with their mothers until they were ready to be weaned at three weeks or so. To our surprise, about half of the pups born to mothers fed starch or white sugar died when they were between 10 and 15 days old, whereas most of those born to mothers fed brown sugar survived until they were weaned at 22 or 23 days. We repeated these experiments several times, until about 300 pups had been born to mothers on each of the three diets. Of the total of 909 pups born, the survival score was 37 percent from mothers fed starch, 53 percent from mothers fed white sugar, and almost 90 percent from mothers fed brown sugar. What is more, every one of the "starch pups" and "white-sugar pups," even those that survived, were clearly ill, with swollen abdomens and weak hind legs; on the other hand none of the "brown-sugar pups" showed these abnormalities.

We were unable to identify whatever it was in the brown sugar that kept the pups alive and well. We did, however, get as far as showing that it was not some complex "biologically active substance," since the effect was still demonstrated when we incinerated the sugar to ash. This burned off all the organic material as well as the sugar itself, leaving only mineral salts. When this ash was added to the mothers' white-sugar diet most of the pups survived, just as they did when the mothers were fed on the brown-sugar diet.

What conclusion can we draw, then, about the comparative value of the white and brown sugars? First, we can be sure that the colored brown sugars have no measurable nutritional advantage over white sugar; even when the only addition is molasses, the quantity is far too small to contribute anything worthwhile. Second, we have not found so far that raw sugar modifies any of the undesirable effects

Typical comparative retail prices of sugar
(Price of granulated taken as 100)

White sugars
consisting entirely of refined white cane or beet sugar

Granulated	**100**
Caster	**130**
Cube	**170**
Preserving	**200**

Raw sugars
consisting of unrefined cane sugar

Golden granulated raw cane	**115**
Demerara raw cane	**140**
Dark muscovado raw cane	**200**

Brown sugars
consisting of refined white cane or beet sugar with added molasses

Light soft brown	**150**
Dark soft brown	**150**

---

of white sugar. But, third, I have to say that the dark muscovado sugar, which carries with it a sizable proportion of the molasses from which it crystallizes, does contain some materials that in some circumstances can contribute to the nutritional value of the diet.

We carried out our experiments not so much because we thought they might tell us something directly about the effect of raw sugar on the health of baby rats, but because the whole process of reproduction—pregnancy, giving birth, and lactation—is a period of physiological stress. A diet that is for most purposes just adequate is more likely to show a marginal nutritional inadequacy when such a physiological stress is imposed.

If, then, I am asked whether one should eat brown sugar or white, my answer is in two parts. First, for reasons that are explained in the rest of this book, I strongly believe that it is better not to eat sugar at all. Second, if you feel that you must take sugar, then it makes sense to eat brown sugar, provided it really is a good quality raw sugar: you should choose a clean, dark muscovado sugar, which contains the greatest proportion of molasses and so of the unidentified nutrients. You should also remember that it is white refined sugar that is used by the manufacturers of all the common soft drinks, ice cream, confectionery, chocolate, and sweet cakes and biscuits.

6

Refined and unrefined

It is popular nowadays to speak of "refined" and "unrefined" foods, and in particular of "refined" and "unrefined" carbohydrates. These terms are most often used in speaking of white sugar and of bread made from white flour. I deplore this custom for two reasons.

The first is that the refining of sugar and of flour are not really comparable. White flour is made by the removal of the bran and germ, and perhaps some of the outer layers of the endosperm, the innermost part of the wheat berry. Everything that has been removed is in fact edible, and would have been eaten if the whole of the berry had been ground. Such flour would contain 100 percent of the wheat berry; what is called wholemeal flour consists of 92 percent of the wheat berry, and white flour usually about 72 percent. On the other hand, the first stage in producing sugar from the cane is the preparation of the cane juice, which leaves behind the major part of the cane as inedible fiber and associated gums and insoluble materials. The following stages of clarification, precipitation, concentration, and crystallization remove further unwanted materials, so that the resulting "unrefined" raw sugar represents only a small proportion of the original cane from which it was produced. This product is far removed from the original sugarcane; the fibrous bagasse from which the cane juice is extracted, and the materials removed from the juice, amount to well over 80 percent of the cane, and what is removed is either inedible or undesirable.

Raw cane sugar thus consists of about 20 percent of the original sugarcane, white cane sugar of perhaps 15 or 16 percent. It does not make sense, therefore, to imply that unrefined sugar is somehow the "whole" or "natural" product of the sugarcane, while refined sugar is in some way "unnatural" or "de-natured." Thus, while the use of these terms, much as I dislike them, may to some extent be justified in regard to wholemeal flour and wholemeal bread, they are invalid where sugar is concerned.

There is a second reason for pleading that you do not speak of refined and unrefined carbohydrate. It is true that refined sugar is the pure carbohydrate sucrose, while raw sugar is mostly this carbohydrate with small quantities of other materials. On the other hand, white flour is not, as some people imagine, virtually nothing but the carbohydrate starch. For example, white flour contains only fractionally less protein than does wholemeal flour—about 13 percent instead of about 13.5 percent. And in many countries such as the USA and the UK some of the vitamins that are partly removed in the milling process are replaced by the flour millers. Moreover, other nutrients are sometimes added to a much higher level than was present in the original wheat grain—for example, calcium in the UK. Altogether, then, it is wrong to call white flour or white bread "refined carbohydrate." And particularly it is not sensible to put on the same nutritional level raw sugar and wholemeal bread, or white sugar and white bread.

Fiber

There are many who consider that the dietary change most relevant to the pattern of disease in Western countries is the change from diets with a high proportion of unrefined foods to diets with a high proportion of refined foods. The evidence for this claim is largely the fact that the diets of people living in rural areas of Africa consist largely of fiber-rich unrefined cereals, and it is in these areas that coronary thrombosis and other diseases of affluence are rare. In the West, where these diseases are common, we have changed from eating brown bread to eating white bread, so that our diet now provides substantially less fiber.

This idea is based on the assumption that cereals are a sizable and "natural" part of the human diet. This is, literally, a short-sighted view: cereals entered our diet less than 10,000 years ago, which is about one half of 1 percent of the period since we emerged as a separate species. Before this, for at least two million years, our ancestors were—like all other species—hunters and gatherers of food. The brief period since the advent of agriculture, which resulted in a diet containing large quantities of starch-rich, high-fiber foods, such as cereals, is far too short for the human species to have completely adapted to such a diet. In other words, there has been far too little time in evolutionary terms for there to have been a significant genetic change toward any adaptation that may have been necessary for such a diet, and if our present-day diet is lower in cereal fiber than that of, say, a hundred or so years ago, then the trend is toward the sort of cereal-free diet eaten by our pre-Neolithic ancestors.

This is one reason why I have not accepted the view that lack of fiber may be responsible for the diseases of affluence. A second reason is that, as we shall see, evidence gained by comparing populations ("population epidemiology") can be very misleading. People in rural Africa or other parts of the Third World live very differently from those in industrialized and urbanized parts of the world. Not only do we take less fiber, but we take more meat, fat, milk, sugar, and a range of other foods; we eat more in total; we are less active physically, smoke more cigarettes, and are more subject to industrial pollution.

Finally, the experiments that have revealed the considerable changes in the body's metabolism that sugar can produce involved comparing diets containing pure starch (or "refined" flour) with diets containing pure sucrose. The many differences in the effects of the two diets could not, therefore, have been due to the presence or absence of fiber, but must have resulted simply from the presence or absence of sugar.

7

Not only sugar is sweet

The most obvious property of sugar is its sweetness, but it has several others: it aids preservation, provides bulk in confectionery, enhances flavor and appearance by caramelizing with heat, gives "mouth feel" to soft drinks, promotes the gelling of jam and marmalade, and provides calories. Alternative sweeteners fall into two groups. One provides sweetness, but virtually none of the other features I have mentioned; the other provides sweetness together with calories and several, if not all, of the other functions of sugar.

Some properties of sugar with examples of use

Sweetness (beverages)

Flavor enhancer (canned vegetables)

Mouth feel or "body" (soft drinks)

Preservation (candied fruits, jam)

Promotes gelling of pectin (jam)

Produces range of textures (confectionery)

Depresses freezing point (ice cream)

Caramelizes (confectionery, crust on bread)

Decoration (icing)

Fermentable (wine)

The caloric sweeteners are either sugars or they are chemically related to one or other of the sugars. Glucose and fructose are the two sugars most commonly used. Glucose, sometimes called dextrose, is made very easily from starch, which, as we saw, is a large molecule made up of glucose units joined together. When starch is treated with acid or alkali, or with appropriate enzymes, it splits into its component glucose units. Much of the glucose in confectionery is used in the form of a syrup, for example corn syrup made from maize starch. It is less sweet than ordinary sugar.

Sucrose, you will recall, consists of a combination of glucose with an equal amount of fructose. Fructose seems to be the part of the sucrose that produces most of the ill effects of sucrose. It is nevertheless often used instead of sucrose for diabetics, because it does not create an immediate need for insulin, as glucose does. One other possible advantage of fructose is that it is nearly twice as sweet as sucrose, so that less, with fewer calories, is needed to produce the same degree of sweetness.

During the past twenty years or so, it has become increasingly practical to produce a mixture of glucose and fructose from starch, which previously could only be made to yield glucose. The process, developed in Japan, depends on the use of an enzyme called glucose-isomerase, which converts the glucose into fructose. By manipulation of the conditions, the proportion of glucose converted to fructose can be varied, producing a mixture with about equal proportions of glucose and fructose (as in invert sugar), or with up to 90 percent fructose. The final product is usually not crystallized from the solution in which it is produced, being transported and used as "high fructose syrup" (HFS). This is now used on quite a large scale, especially in the USA and Japan, as an alternative to ordinary sugar. Manufacturers can switch from sucrose to HFS and back according to the fluctuations in sugar and starch prices. Because of this, the farmers in Europe who produce sugar beet have persuaded the EEC authorities to put a levy on HFS production and a quota on its import.

The nonsugar caloric sweeteners are made from sugars, and are what chemists call "polyols." By the process of chemical reduction—adding hydrogen atoms, for example to fructose, so that one more alcohol group is formed and added to the five already present—sorbitol is produced. Other polyols that can be used include maltitol and xylitol. They all provide, in a given quantity, about the same number of calories as ordinary sugar (sucrose); however, as they are not as sweet, you would tend to use more, and so take in more calories. These caloric sweeteners are therefore of no help in a slimming diet, but sorbitol is sometimes recommended as an alternative to ordinary sugar for diabetics, and xylitol has been used in candy and chewing gum because it does not harm the teeth. A major disadvantage in these polyols is that, unless taken in fairly small quantities, they tend to cause diarrhea.

The noncaloric sweeteners have no chemical relationship to the sugars and are very much sweeter, so that only tiny quantities are used. For this reason they are sometimes called "intense sweeteners." They have mostly been discovered accidentally in research laboratories where chemists were synthesizing new chemical substances for quite other purposes. They can be used to assist weight loss through reduced calorie intake; to help sufferers from diseases such as diabetes which affect sugar metabolism; to substitute for sugar in times of shortage, for example during war—or, increasingly, I am glad to say, to help to prevent the sugar-promoted diseases described in this book. The best-known noncaloric sweetener is saccharin, which was discovered in 1879. Its use increased considerably during the sugar shortage of the First World War. Another widely used noncaloric sweetener is cyclamate, discovered in 1937; it is, however, at present not used in the USA or the UK. A new, increasingly popular sweetener is aspartame. Because the noncaloric sweeteners do not have the properties of sucrose in providing bulk, preservative power, and so on, they are used almost exclusively as so-called tabletop sweeteners, to be added to tea or coffee, or else in the manufacture of low-calorie cold drinks. To a small extent they may also be used in home cooking, in the preparation of some items such as fruit salad. Their lack of bulk, however, rules them out as sugar substitutes in most desserts, confectionery, and ice cream.

From time to time suspicions arise that the alternative sweeteners could be harmful. This happens most often in relation to the noncaloric sweeteners, presumably on the grounds that most have a chemical composition quite different from that of any of the natural sugars, or, indeed, of any naturally occurring substance. The suspicions usually arise from some superficial or incomplete research that does no more than hint at some possible harmful effect detected in, for instance, laboratory rats which have been fed the sweetener. The result is usually great contention and the setting up of a much more extensive investigation. In these circumstances, tests are often carried out with phenomenally large doses of the sweetener; a recent test with saccharin used quantities that in a human being would require the daily consumption of, for instance, several hundred cans of soft drinks sweetened solely with saccharin. This much saccharin would have the sweetening power of some 5 kg of sugar a day.

It is worth spending a moment here on the question of the toxicity of substances that accidentally or intentionally may find their way into our food. The most important fact to remember is that there is really no such thing as something being poisonous, or something not being poisonous, just like that. What matters is not only the nature of the substance, but also its quantity. No substance is intrinsically harmless; you can make yourself dangerously ill by taking large quantities of water. No substance is intrinsically harmful; it was fashionable in the early part of this century to give medicines containing arsenic as a tonic, although of course the quantities were very small indeed.

Similarly, if it turns out that cyclamate, or saccharin, or anything else, causes some undesirable effect in daily amounts that are fifty or one hundred times as much as anyone could possibly take—and even then only when taken over a period of ten years or more—it would not be sensible to ban it automatically.

In the USA the situation was complicated by what is known as the Delaney Clause, agreed by the U.S. Senate in 1958, which says that "no additive shall be deemed safe if it is found to induce cancer when ingested by man or animal." This has been interpreted as forbidding the use as a food additive of any substance that, in any quantity, and over however long a period, produces cancer in any species of animal. It was this provision that led to the banning of cyclamate in the USA in 1970. This decision was based on the result of one experiment in which a small proportion of rats fed for a long time with very large doses of a mixture of cyclamate and saccharin developed cancer of the bladder. Within a week or two of the American decision, the UK followed its example, so that cyclamate is not used by the food industry in either country, although the position is under review. However, 16 out of the 17 countries of Western Europe do permit the use of cyclamate.

Those people who are still concerned about the possible hazards of taking artificial sweeteners could reduce or abolish their cause for worry by using mixed sweeteners. This should reduce the possibility of being harmed by any one of them, since each would be present in a lower concentration than if it were the sole sweetening agent.

Relative sweetness of sweetening agents
(Threshold sweetness of sucrose = 1.0)

Caloric sweeteners		Noncaloric sweeteners	
Glucose	0.5	Cyclamate	30
Sorbitol	0.5	Acesulfame-K	150
Mannitol	0.7	Aspartame	200
Xylitol	1.0	Saccharin	300
Fructose	1.7	Thaumatin	3,000

At present, the better-known noncaloric sweeteners permitted in one or more of the countries that control food additives are saccharin, cyclamate, aspartame, acesulfame-K, and thaumatin (talin). Their relative sweetness compared with sugar is given in the table. For a variety of reasons, however, these figures are only approximate. First, people's subjective assessment of sweetness varies.

Second, the intensity of some sweeteners increases or decreases with the acidity of the food or drink to which they are added. Third, sometimes their relative sweetness changes with the degree of their dilution and the temperature of the food or drink.

The noncaloric sweeteners are not entirely interchangeable. For example, saccharin and, to a lesser extent, aspartame are not stable to heat, so they are not used in the preparation of dishes that require prolonged cooking.

In addition aspartame, being a compound of two amino acids, aspartic acid and phenylalanine, may cause upset in children born with the condition of phenylketonuria (PKU). Such children are unable to deal with more than a small quantity at a time of phenylalanine, one of the amino acids found in most proteins. If more than this limited amount is taken regularly, a substance is produced that can cause mental impairment. Children usually grow out of PKU by the age of 10 years or so. Meanwhile the condition is controlled by giving the sufferer a carefully constructed diet containing sorts and quantities of protein that enable the phenylalanine intake to be limited. In addition, a child with PKU should be made aware of which soft drinks are sweetened with aspartame, and be taught to avoid these.

8

Who eats sugar, and how much?

People look at me quite incredulously when I tell them that there are now many parts of the world where the average person— man, woman, and child—is eating more than 100 pounds of sugar a year—two pounds or more a week. But though this is true today, it has only rather recently become so and it still isn't true for all countries. In this chapter I want to tell you how sugar consumption has been changing, how much is being eaten in different countries and by people of different ages, and how much of Western man's consumption is ingested by way of different sorts of manufactured foods and drinks, along with the sugar to which people help themselves from the bowl at the table.

Before going any further, I should emphasize that in this book I am talking about the sugar (sucrose) produced from the cane and beet. This is technically called centrifugal sugar. I am excluding sucrose produced from other sources such as the maple and the palm; the amounts are negligible and come to only 1 percent or so of the total. I am also excluding milk sugar (lactose), as well as the sucrose and other sugars one consumes in fruit and vegetables. The reason here is also chiefly quantitative; the amounts of centrifugal sugar are much greater than those of the sucrose from other sources. In one of our studies, we found that adults ate about half of their total carbohydrate as starch, 35 percent as centrifugal sucrose, 7 percent as lactose, and the remaining 8 percent or so as the mixed sugars in fruits and vegetables—mostly glucose, fructose, and sucrose.

In the year 1850 world production of sugar was about 1½ million tons. Forty years later it was more than 5 million tons, and by the turn of the century it was more than 11 million tons. Except for a setback during each of the two world wars, production has continued to rise rapidly, so that it reached 35 million tons by 1950 and is now more than 100 million tons. Over the past 100 years there has been a 25-fold increase in world sugar production; allowing for the increase in world population, this represents an increase in average consumption from 7 pounds a year to 45 pounds. The most extensive statistics of sugar production and consumption were collected 25 years ago in a report produced for the Food and Agriculture Organization of the United Nations. Although this is now a little out of date, I shall quote some of its findings because they still demonstrate many interesting features that are not easy to discover from more recent statistics.

World sugar production

	million tonnes
1800	0.25
1850	1.5
1880	3.8
1890	5.2
1900	11
1950	35
1970	70
1982	101

During the 20 years from 1938 to 1958, there was an increase in world production of many commodities. Among food items, cocoa increased by 20 percent, milk by about 30 percent, meat and food grains up to 50 percent, but sugar production outstripped all of these with its enormous increase: 100 percent over the 20 years. Between 1900 and 1957, consumption of sugar increased from an average of 11 pounds a year to 34 pounds; by now, as I said, it is about 45 pounds. But the increase has differed in different countries. It has been most rapid in the countries that until recently had a low consumption.

Before the last war, Italy's yearly average was less than 20 pounds; by 1970 it was more than 60 pounds. Those countries that already had a high consumption have had a smaller increase or none at all; in the United Kingdom there was an increase from about 100 pounds to 120 pounds, while in the United States there has been no change from the previous 102 pounds or so. It looks as if there is a limit of somewhat over 100 pounds a head a year at which all countries stop increasing their intake. The wealthier countries gradually achieved this high level by a slow and fairly steady increase over perhaps 200 years; some of the poorer countries are now achieving it very much more rapidly.

The best statistics for any one country over a long period are those for the UK. Just over 200 years ago we used to take 4 or 5 pounds of sugar (about 2 kg) a year; by the middle of the nineteenth century, this had increased fivefold to about 25 pounds a year; we now take about 100 pounds a year. Over the whole 200 years we have increased our consumption 25-fold. To put this another way, 200 or so years ago we used to spend a whole year getting through the amount of sugar we now get through in two weeks.

The apparent fall in the consumption of centrifugal sugar in the UK after about 1970 is almost exactly equaled by an increased consumption, largely in manufactured foods, of glucose and of small quantities of the recently introduced High Fructose Syrup. Total consumption of all three forms of sugar has hardly changed in the past 20 years or more.

There are also some figures for other countries or populations. In Switzerland, average intake has increased tenfold in the last 100 years. Consumption among Canadian Eskimos increased much more rapidly; in one area, it rose from 26 pounds to 104 pounds a

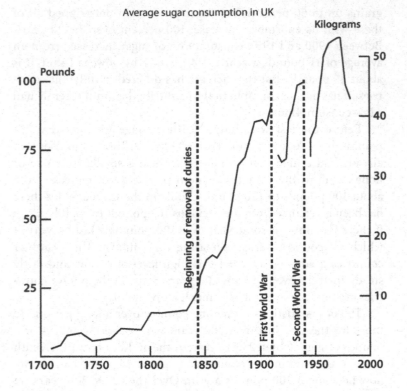

Average sugar consumption in UK

year between 1959 and 1967. The consumption among the rural Zulu population in South Africa increased tenfold in eleven years, from six pounds a year to 60 pounds a year between 1953 and 1964.

We have been looking at the way sugar consumption has been going up especially during the last 200 years or so, and also at the way sugar consumption differs in different countries—generally high in wealthy countries and low in poor countries. I should like to say a little more about diets in rich and poor countries because, although not directly related to sugar consumption, they do have a bearing on this, and they also give us a better picture of the way diets are affected by income. Let us look at the diets in different countries according to their average national income, and calculate

German sugar consumption (per head)
(West Germany only from 1950)

	kg/year
1825	2
1850	3
1880	8
1914	18
1939	26
1960	30
1970	34
1980	36

how many calories were supplied by these diets, how much protein, fat, and carbohydrate, and how much of the carbohydrate was made up of sugar on the one hand and of other components—chiefly starch—on the other hand.

As you pass from the poorest to the wealthiest group of countries, you find an increase of about 50 percent in the average number of calories in the diet, from about 2,000 calories a day to about 3,000. Protein increases by about 80 percent from 50 grams to 90 grams a day, and fat increases about fourfold from 35 grams to 140 grams. The total amount of carbohydrate is much the same irrespective of wealth, except that it is somewhat lower in both the very poorest and very wealthiest countries. In the very poorest, people just have too little of everything. In the very wealthiest countries, consumption of foods rich in protein and fat is high enough to cause a small reduction of foods rich in carbohydrate.

But more interesting than the general similarity in total carbohydrate is the very considerable change in the sorts of carbohydrate as you pass from poor to rich countries. There is a great increase in the amount of sugar, and a corresponding fall in the remaining carbohydrate, mostly starch. This is similar to the situation when a

Countries with lowest and highest sugar consumption
(per head, 1982)

Lowest consumption		kg/year		
	Kampuchea	0.7		
	Rwanda	0.8		
	Ghana	1.2		
	Nepal	1.2		
	Uganda	1.4		
	Laos	1.5		
	Burma	1.7		

Highest consumption	Major cane sugar producers		Other countries	kg/year
	Cuba	66.6	Iceland	52.2
	Costa Rica	62.9	Israel	52.2
	Fiji	60.3	Hungary	52.1
	Barbados	60.2	New Zealand	50.9
	Guyana	51.9	Singapore	50.6
	Australia	51.6	Austria	50.5

particular country becomes increasingly wealthy: more sugar is eaten—and less bread, rice, corn, potatoes, or other starchy food.

The figures I have given so far are averages for whole populations. When I tell an audience in London that they eat 5 ounces of sugar a day, they profess astonished disbelief. Everyone insists that they eat much less than this, so I usually say that since 5 ounces is the daily average there must be other people who are eating more.

A frequent criticism of the experiments (to be described later) carried out in Queen Elizabeth College Nutrition Department is that we use excessively large amounts of sugar; the apparent ill effects

UK consumption of soft drinks (per head)

	gals/year
1939	2.7
1950	4.1
1960	7.9
1970	10.7
1980	18.2

produced would not, it is said, be caused by the amounts that ordinary people consume. After all, it is argued, immoderate quantities of any food might be harmful. When we first reported that sugar in the diet increases the amounts of the fatty substances in the blood (notably triglyceride and cholesterol), an American scientist wrote that sugar produces *no* increase in triglyceride if the amounts taken "are of the same order of magnitude as the average sugar intake of the American population." Similarly, another researcher claimed that "there is little relationship under ordinary conditions between dietary sugar and plasma cholesterol." These references to "average" and "ordinary conditions" assume that virtually everybody takes an amount of sugar that is little different from the average intake, which in America and in the UK is about 125 grams a day. This is about as sensible as saying that everyone takes about an average amount of alcohol, so that alcohol cannot be a cause of liver cirrhosis.

U.S. consumption of soft drinks (per head)

	bottles/year
1950	40
1960	190
1980	300

Though there is little published information about sugar consumption in individuals, common experience tells us that it varies a great deal. There are people who take no sugar in their tea or coffee, rarely if ever take sweetened drinks, eat little confectionery, and do not ordinarily eat desserts. There are others who begin the day with sweetened cereal and added sugar, have sugar in all their hot beverages, eat sizable quantities of confectionery, cakes, and biscuits between meals, and always have a cooked and heavily sweetened dessert with their main meals. The meager figures on individual sugar intake that have been published confirm that there is a considerable range. In our Queen Elizabeth College studies, we measured how much sugar was being taken by various groups of older children and by men and women of different ages. They are not necessarily representative, but I give you our results in the table below because they demonstrate some general features.

Let me add that it is likely that we are underestimating the exact consumption, because people tend to forget the occasional sugar drink or piece of chocolate they have been taking. Still, one can get some interesting information even if it is somewhat approximate.

Daily sugar intake (per head, in grams)

Age	males	females
15–19	156	96
20–29	112	101
30–39	126	100
40–49	96	83
50–59	90	83
60–69	92	63

The most striking feature is the very high consumption by teenage boys; more than 50 percent above that by teenage girls. The sex difference persists throughout later life although not so strikingly. From the age of 20, men take something like 15 or 20 percent more sugar than do women. This is possibly because women are more weight-conscious, so they deliberately—and wisely—restrict their sugar consumption. A decline of sugar intake sets in with increasing age, so that people in their sixties take about one third less sugar than do people in their twenties.

These figures come from our own studies in London, but I have also tried to find statistics reported by others. Mostly, however, these cover only some sugar items.

A study of over 1,000 American boys and girls aged between 14 and 18 in the state of Iowa showed an average sugar consumption by the boys of 389 grams a day and by the girls of 276 grams. This amounts to an average of more than 40 percent of their total calories as sugar; the average of the whole population in the USA was something like 18 percent. In a study of 17-year-old white children in South Africa, sugar consumption was not as high, yet one third of the boys took an average of 241 grams and one third of the girls an average of 171 grams.

National chocolate consumption (per head, 1980)

	kg/year
Switzerland	7.2
W. Germany	6.6
Netherlands	5.0
USA	3.9
Italy	1.0

In Scotland, dentists examined 13-year-old boys and girls, a younger age group than any we have studied. They estimated only the amount of confectionery the children ate, and they added that they were sure that their figures were in fact underestimates. The average weekly intake was 17½ ounces, boys eating slightly more than girls. Eight percent of the children, however, took more than 32 ounces a week.

National ice cream consumption (per head, 1982)

	liters/year
USA	25
Sweden	12.5
Switzerland	7.9
UK	5.2
Italy	5.1

These figures are equivalent to a daily intake of about 2 ounces (55 grams) a day of sugar from confectionery for all the children, and nearly 4 ounces (105 grams) for 8 percent of the children. The average intake of confectionery for the whole British population is 8 ounces a week, which is matched only by Switzerland. According to figures published by the British confectionery industry, consumption in children under 16 is about 17 ounces a week, roughly the same as that reported by the Scottish dentists.

As well as confectionery, of course, teenagers undoubtedly eat more than adults do of such items as cakes, biscuits, ice cream, and desserts. Even at a conservative estimate, these can be expected to bring the total amount of sugar to something like 50 percent more than the national average. This would make the total consumption of a 13-year-old about 7½ ounces of sugar a day, which would supply 850 calories out of their daily total intake of about 3,000 calories. Now think of the children who eat not 17½ ounces of sweets a week but more than 32 ounces, and it is pretty certain that there must be a lot of children getting at least half of their calories from sugar.

You might perhaps think that, eating a lot of sugar between meals, they would cut down the sugar in meals. Not at all. A colleague of mine found that the midday dinners in several English schools contained sugar giving about 25 percent of the calories, and on the whole children get the same sort of food at school as at home. So it does look as if children get more than the average amount of sugar, sometimes much more; not only in the snacks and drinks between meals, but also in the meals themselves. Part of this, I am sure, is due to the attitudes of their parents, who wish to give pleasure to their children, to win their affection, and to provide them, as they believe, with the energy they need for growth and work and play.

The London *Times* reported the case of a young lad eating more than 6½ pounds of sugar a week, which amounts to nearly 350 pounds in a year. His dentist complained that six months after his mouth had been made quite free from decay, it was now once more full of rotting teeth. Our own record comes from a 15-year-old boy who also consumed just under a pound of sugar a day, or around 1,700 calories from sugar alone.

Of course, just as there are some people who eat very much more than the average amount of sugar, there must be those who eat less than the average. Our own figures suggest that the range of variation of sugar intake is far greater than the range for most other foods. We have found people taking as little as half an ounce a day (15 grams) as well as those taking as much as 14 ounces a day (400 grams); the latter are eating in one day what the former eat in a month.

Altogether, I find it difficult to resist the conclusion that, whereas the national average consumption of sugar in the U.S. and the UK represents something like 17 percent or 18 percent of the average calorie intake, the average for children would work out at around 25 percent of the calories or even a little more. And again let me say that there must be some who are getting 50 percent of their calories from sugar. In absolute terms sugar consumption for many children must amount to nearer 10 ounces a day than the 5-ounce national average.

In case you think that I am exaggerating the amount of sugar taken by children, let me quote from an advertisement by Sugar Information, the public relations organization for the American sugar industry. Forget for now the reference to obesity. I shall be saying something more about this aspect of sugar later. Here is part of the advertisement:

> You've probably had people tell you they're avoiding this or
> that because it has sugar in it. If you want to see how much

sense there is to that idea, next time you pass a bunch of kids, take a look. Kids eat and drink more things made with sugar than anybody. But how many fat ones do you see?

Good nutrition comes from a balanced diet. One that provides the right amounts, and right kinds, of proteins, vitamins, minerals, fats and carbohydrates. Sugar is an important carbohydrate. In moderation, sugar has a place in a balanced diet.

The word I like best in this advertisement is "moderation." But would you really accept as moderate the current average consumption of sugar by kids, probably amounting to 25 percent or more of their calories and adding up to 7 ounces or so a day?

Let me pursue this concept of moderation, about which we hear so often. Supposing we were living a couple of hundred years ago. People in America and Britain were then eating on average a couple of ounces of sugar a week. If someone were then to have said that you should eat sugar in moderation, you would have thought in terms of perhaps no more than 3 ounces a week. You would certainly have protested that 7 ounces a week—an ounce a day—was a quite excessive amount. But people today accept *five* ounces a day as moderate; only when someone eats much more than this does it become generally accepted that they are eating immoderately.

Look now at babies, who are bottle-fed more and more, even though there is a slight drift back to breast-feeding in some middle-class homes. A common feeding formula consists of dried cow's milk, perhaps modified in some way, with added sugar. Except in some sensible preparations, the sugar that is added is sucrose, not lactose (milk sugar), and I shall show later that this is not at all the same in its effect on the baby. Here I refer only to the disadvantage of sugar having a much sweeter taste than lactose, so that a baby is inducted into his later sugar-rich life by being encouraged to develop a taste for maximum sweetness.

As soon as a baby begins to receive mixed feeding—and this is often at two or three months or even earlier—cereal will be added to the diet, and then foods like egg yolk and minced meat and sieved vegetables and fruit. Many mothers will add sugar to the cereal and to the fruit, although it is by no means uncommon to add it also to egg and meat and fish. And I have not mentioned the pernicious habit of giving babies dummies that have a reservoir for syrup or which from time to time are dipped into the sugar bowl.

I knew of a family of four people: father, mother, a girl of 4, and a baby of six months. They buy and use 11 pounds of sugar a week, and this does not prevent them from also buying the usual assortment of biscuits and ice creams and other manufactured foods and drinks with sugar. The baby certainly gets less than a quarter of all this, but it is hardly deprived since its dummy is dipped into the family sugar bowl.

UK industrial use of sugar (approximate, 1980–81)

	thousand tonnes
Chocolate and confectionery	320
Biscuits and cakes	250
Soft drinks	250
Ice cream and milk drinks	85
Canned and frozen foods	65
Jams and preserves	60
Pharmaceuticals	25
Miscellaneous	25
Brewing	45

One of the reasons why some people find it difficult to accept that on average Americans and Britons eat about two pounds of sugar a week is because they think only of the sugar that is brought into the home as visible sugar. But an increasing proportion of sugar is now bought already made up into foods. If you look at your own sugar consumption, the chances are that over the years a smaller and smaller fraction will be household sugar and a greater and greater fraction industrial sugar. Household sugar is mostly what is bought by the housewife, but also includes the much smaller quantity used in cafés and restaurants. Industrial sugar goes to the factory and comes to us in the form of confectionery, ice cream, soft drinks, cakes, biscuits, and nowadays also a very wide range of other items, especially the fancily packaged "convenience foods."

The poorer countries, as you might expect, consume less of their sugar in the industrial form; manufactured foods are a luxury consumed increasingly in the wealthier countries. In the late fifties, according to the FAO report I mentioned earlier, South Africa took only 20 percent of its sugar in manufactured foods, while France took 40 percent and Australia 55 percent. American manufactured sugar increased from less than 30 percent in 1927 to about 50 percent in 1957, and is now more than 70 percent. The increase in the proportion of manufactured sugar in the USA is especially interesting in view of the fact that the total sugar consumption there has not changed much over this period.

Proportions of domestic and industrial use of sugar

		Domestic %	Industrial %
West Germany	1957–58	55	55
	1981–82	30	70
USA	1934	70	30
	1944	57	43
	1954	44	56
	1964	38	62
	1974	34	66
	1983*	39	61

* Between 1974 and 1983 the use of High Fructose Syrup increased from 3 to 43 pounds a head; almost all of this was used in food manufacture as an alternative to sugar. If this is taken into the calculation, industrial use of sugar in 1983 was 74% and domestic use 26%.

The UK use of manufactured sugar amounts to about 65 percent. The ways in which sugar is used by food manufacturers in the UK, and the various amounts involved, are shown in the table on p. 53. But I want to amplify these figures in several ways. To begin with, I believe there are several reasons why Westerners continue to increase their consumption of manufactured foods containing sugar. One is that any efficient manufacturer is constantly producing more and more attractive foods. Because of competition, he keeps making new products or new variations on his old products, each time with the purpose of producing something that is even more attractive than before. More and more, people find it difficult to resist these delicious foods and drinks. In 1981, nearly £100 million was spent on advertising sugar-rich foods; £53 million of this was spent on advertising chocolate and confectionery.

Secondly, sugar, as we have seen, offers many more properties than just sweetness. Its use in different sorts of confectionery depends also on its bulk, on its ability to exist either in crystallized or non-crystallized form, on its solubility in water, and on its change of color and flavor when heated. Its use in jams depends on its ability to set in the presence of pectin, and on its high osmotic pressure, which inhibits the growth of molds and bacteria. In small quantities, sugar seems to enhance the flavor of other foods without necessarily adding specifically to sweetness. These and many other properties of sugar amount to an extraordinary versatility, and account for its use in such a vast range of foods and drinks.

The result is plain to see if you walk around the supermarkets and make a list of foods with sugar among their ingredients. Leaving aside obvious items like cakes, biscuits, desserts, and soft drinks, you will find sugar in almost every variety of canned soups, in cans of baked beans and pastas, many kinds of canned meat, almost every breakfast food, several frozen vegetables and made-up dishes, and most canned vegetables. In some of these foods, especially in the foods like meats or vegetarian meat substitutes, the amounts of sugar are quite small. But in many others the amount is really surprisingly high. You can get some idea by seeing where sugar ranks in the list of ingredients. If it is first in the list, the food contains more sugar than any other ingredient. When I tried this exercise, this was true of one or two canned soups, one or two breakfast foods, and several pickles and sauces.

A third reason why people increasingly buy manufactured foods containing sugar is that they prefer to buy foods in "convenience" form—usually items that they would previously have made for themselves. And it looks from my sampling as if these foods are likely to contain more sugar than they would have done when made at home. The manufacturer seems to have found, or at any rate convinced himself, that people like sugar with everything, and more and more of it. In the last two or three years I have found it difficult at a bar to get tomato juice—my favorite tipple—that has not had sugar added to it. I am also rather fond of peanut butter, but the manufacturers of the two most popular brands in England have now decided that I ought to have it with sugar. Here let me give one good mark to the health food people; at least some do not put sugar into the peanut butter—anyway, not yet.

If you want to test what I am saying, try next time you are out to get a drink of something or other that is nonalcoholic, does not contain sugar, and is not specially advertised as a "diet drink."

It does seem to be true that until they reach a certain limit most people demand more and more sugar as they go on taking it. Certainly the converse is true. Many people have been restricting sugar for some time, either because they are concerned about their weight or for even more serious reasons; now, when for social reasons they do have to take sugary foods and drinks, they often find them intolerably sweet. On his third birthday, my well-brought-up grandson Benjamin took one bite of his iced birthday cake and ate no more because, he said, "It's too sweet."

What is surprising to me is the high proportion of sugar in many so-called health foods besides the peanut butter I have mentioned. Sugar appears to figure prominently in foods that are supposed to be "good for you." Eggs and bacon, or the old British favorite kipper, would be better for you than several of the special breakfast health foods such as many brands of muesli.

One more reason why Westerners eat so much sugar is that

increasing affluence gives people more leisure, creating the kind of situation—sitting in front of the television, making a trip in the car—that is conducive to the consumption of snacks and soft drinks, so easily available nowadays, and considered to be inexpensive. And snacks usually, and soft drinks almost always, are rich sources of sugar.

Another point about soft drinks. When I was young, if I was thirsty I had a glass of water. Nowadays when children are thirsty it seems almost obligatory that they quench their thirst with some sugar-laden cola or other drink. And this is often true for adults too, although it is just as likely to be an alcoholic drink like beer. In this way, sugar is consumed almost inadvertently. The modern trend of using drinks like tonic water or bitter lemon as mixers is for many people a further source of sugar of which they are hardly aware. Two small bottles with your gin or vodka and you have swallowed an ounce or more of sugar.

Life is difficult for people who, like myself, want to avoid sugar, and particularly for those who, like the people with hereditary fructose intolerance, get sick when they take sugar. But I am glad to see that an increasing number of manufacturers put no sugar into some of their products, and that you can find more and more labels marked "sugar-free" or "no added sugar." In particular it is encouraging to see more baby foods labeled in this way.

9

Words mean what you want them to mean

It is very confusing when people use different words for the same thing. In England, we say "lift" for what the Americans call "elevator," "property" when they say "real estate," and "petrol" when they say "gas." But even greater misunderstandings arise when people use the same word for different things. The American woman carries a handbag which she sometimes calls her purse, while an English woman carries a handbag in which she has a much smaller purse for her money. The American woman carries her money in her wallet.

As we saw in Chapter 3, "sugar" sometimes means the beautiful white powder or lumps that this book is all about—sucrose—but sometimes it means a different substance that circulates in the blood—glucose. Another example is the word "energy," which, as I shall discuss, means one thing to the nonscientist and quite another to the nutritionist.

Glucose is a sugar that is found, usually with other sugars, in some fruits and vegetables. It is very important to biochemists, physiologists, and nutritionists because it is a key material in the metabolism of all plants and animals. Many of our principal foods are sooner or later converted into glucose, and glucose forms one of the most important substances that is metabolized (or oxidized or burned) in the tissues to supply energy for everyday activities.

Where energy comes from

Almost every book written by people in or associated with the sugar industry contains a section in which you are told how important sugar is because it is an essential component of the body. They tell you that it is oxidized so as to give energy, that it is a key material in all sorts of metabolic processes, and so on. And they imply or even say explicitly that all this is to do with "sugar" (sucrose), whereas in fact they have been talking about "blood sugar" (glucose). The fact is that sucrose and glucose have different chemical structures and their effects in the body differ in important ways. When the word "sugar" is used at one moment to mean the sucrose in your food and at another to mean the glucose in your blood, these differences are hidden. So accustomed do we then become to this blurring of definitions that eventually we find it difficult to accept the vital differences between the sucrose we eat and the glucose in our blood.

There is a second way in which you may be led to believe that sugar is an important, if not essential, item of our diet. Here is a quotation from a pamphlet from the sugar industry: "Sugar works for you with each bite you eat—for your body is an energy factory with sugar as its fuel." Firstly, it is not "sugar" (sucrose) but "sugar" (glucose) that is the body's fuel, and secondly, what does "energy" really mean? When you say, "I have no energy," or "Little Johnny is full of energy," you use the word to mean either physical activity or the inclination to be physically active. When you say that Johnny is full of energy, you picture him rushing around, leaping up and down stairs, climbing a tree, or tearing along on his bicycle. On the other hand, when you say you have no energy you imply that you do not want to do anything much other than sit about, or preferably lie down.

So when someone says, "Sugar gives you energy," you imagine that this is just what you need to leap out of your chair and dash around like little Johnny. But the physiologist and the nutritionist who talk about sugar and energy mean something different. What they mean is that sugar (like any other food, after having been digested and absorbed) can be utilized by the body so as to release the energy you need for all the functions of the body. These include such automatic activities as breathing, heartbeat, or digestion, and all the chemical reactions of the living body that add up to what is called "metabolism." They also include such voluntary activities as dressing or walking or running.

What people really mean when they say that sugar gives them energy is simply that it is a potential source of the energy needed for the processes of living. It is there when you need it, in the same sort of way as the petrol (or gas!) that you put into your car is in the tank, ready to be burned when you want the car engine to go. Just putting another gallon or two in the tank does not, of itself, make the car go any faster or make it any more energetic. And taking another spoonful of sugar does not, of itself, make you jump out of your chair and rush to mow the lawn.

All food then contains "energy," in that some of its components can provide the fuel for the body's workings. Normally you have quite a sizable reserve of this fuel in your tissues, stored from the food you have eaten on previous occasions. If you were starving, so that you had little or none of this reserve, and if in addition it were imperative that you have some fuel in your tissues within minutes, in addition to the glucose in your blood, then it might be a good idea to eat sugar rather than any other food because the sugar quickly gets digested and absorbed and taken to the tissues. A piece of bread and butter would take a few minutes longer. This insignificant time differential is what the sugar propagandists mean when they talk about sugar's "quick" energy. But isn't it really quite rare for circumstances to arise that make it imperative for you to take advantage of this more rapid availability of "energy" from sugar? And besides, as we shall see later, it may be that the rapidity with which sugar floods the bloodstream is harmful rather than beneficial.

I sometimes wonder whether the insistence that sugar contains energy arises from the fact that it contains nothing else. All other foods contain energy as well as at least *some* nutrients in the way of protein or minerals or vitamins or a mixture of these. Sugar contains energy, and that is all.

Pure is good

As I have shown, the combination of all foods contains the whole range of essential materials that the body needs for its survival and well-being. Each one of these is derived from living plants or living animals; if they are not processed in any way, they contain a mixture of approximately 50 essential materials. From a cabbage, you obtain among other essentials some vitamin A and vitamin C and calcium. From a piece of meat you obtain protein, fat, several vitamins of the B group, iron, and many other nutrients.

But suppose one were to cultivate pine trees instead of cabbages, and then extract the vitamin C and eat that instead of eating cabbage? It would be possible now to claim that you have consumed absolutely pure vitamin C, but it would not be of any particular advantage to get it this way rather than from the cabbage. In fact you would lose out in this transaction because the cabbage would have given you other nutritional benefits apart from vitamin C.

Yet this is really the sort of thing people do when they make sugar. They plant vast areas of land with sugarcane or sugar beet instead of crops that they can eat more or less whole. Then they take the cane or beet and extract, clean, filter, refine, and purify it until they have something that is virtually 100 percent sugar. At this point, the refiners say with absolute truth that this sugar is one of the purest foods known.

Once more a word is being used in two different senses. When you say water is pure, or bread, or butter, you mean that it is not contaminated with anything inferior, and especially not contaminated with anything harmful. But then you are persuaded to carry over this sense of wholesomeness to the chemists' meaning: a material that does not have something else mixed with it, irrespective of whether this something would have been harmful or harmless or even beneficial.

There is no special reason to praise sugar for the fact that, in the course of its elaborate preparation, it is freed from all other materials so that it is chemically "pure," as are most of the other materials the chemist has on his laboratory shelves. Equally I would see no reason for being pleased at being presented with pure protein for my consumption, or pure vitamin B_{12}, or any other dietary component in its isolated state. What virtue would this represent?

10

Sugar's calories make you thin—they say

The inclusion of large amounts of sugar can affect our diet in two ways. It can be taken *in addition* to the normal diet, or *instead* of a calorie equivalent in some other food. More likely than either of these alone, it can be done both ways: by adding something to the total calories, and also displacing some other foods. Since, as I showed, sugar supplies nearly one fifth of the average eater's calories, no aspect of sugar consumption can be ignored. Its effects will be particularly evident in those many people whose intake of sugar is appreciably greater than the average.

The consumption of sugar on top of an ordinary diet increases the risk of obesity; the consumption of sugar instead of part of an adequate diet increases the risk of nutritional deficiencies. In this chapter, I want to deal with the question of sugar consumption leading to an increase in calorie intake.

I have already pointed out that the average intake of sugar in America or Britain supplies some 500 to 550 calories a day. But this is not the whole picture. Many people take at least twice as much as the average of 4½ or 5 ounces a day; they are getting at least 1,000 calories a day from sugar, and 1,500 calories or even more is not unknown. This sounds enormous, but I am not counting visible sugar alone. Such people consume only part of this daily quota as sugar by itself. Moreover, much is taken with other foods that supply lots of calories: cocoa in chocolate, fat in ice cream, fat and flour in biscuits and cakes. This adds up to even more calories than the figures I have just given.

This book is not about obesity and its causes and treatment, so I shall mention only two matters that are particularly relevant to the question of sugar—one obvious, one less obvious and only recently properly investigated. The obvious one is that people take sweet foods and drinks because they like them. And just as you will eat less than you need if your food is unpalatable and unappetizing, so you will eat more than you need if it is especially appetizing.

Let me remind you of some of the points I made in Chapter 2. Most often, people eat chocolate or cake because they are tempted by their appearance and taste, and not because they really need those extra calories. And when people take sugary soft drinks, they usually do so because they are thirsty rather than because they are hungry, even though the drinks supply lots of calories (probably not needed) along with the water that *is* needed. Thus, people often eat and drink to satisfy appetite—for pleasure, rather than to satisfy hunger.

It is worth spending a moment or two more on this distinction between appetite and hunger. What are the foods that make overweight people overeat? Mostly, people don't become overweight because they eat too much meat or fish, or too many eggs, or too much fruit or vegetables. It is almost always that they eat too much bread, or sweets and chocolates, or cakes and biscuits, or because they drink too many sugary cups of tea or soft drinks. Or, of course, it may be because they drink too much beer or other alcoholic drink.

Now just think. When people put sugar in their tea or coffee, is it because they are hungry and need the extra calories? Or is it that they prefer the beverage sweet? If it were really a question of caloric needs, then they would be adding the sugar only when they were hungry.

Or take someone who goes to the pub after his supper and drinks two or three pints of beer with his friends. Is this because he is short of calories? Does he go to the pub only when he is hungry? Or does he drink just half a pint during the evening when he has had a particularly large meal at home?

And what about the woman who sits in front of the television after supper, with a box of chocolates on her lap. Does she eat only one chocolate because she had a large dinner that night, compared with the half-box of chocolates she ate the previous night when she was really hungry? The fact is that on both occasions she nibbles chocolates because she likes them, and this has nothing whatever to do with her hunger.

In general, people take sugar or sugary foods or drinks or alcohol for pleasure. The calories they inevitably get at the same time are quite incidental and have nothing to do with the satisfaction they get from consuming these items.

When you come to think of it, almost all of the tempting foods that are taken to satisfy appetite rather than hunger contain carbohydrate that is either sugar or starch, or they contain alcohol. This was confirmed when my colleague Diane Adie and I carried out a survey among more than 1,400 women who were members of *Slimming Magazine*'s Slimming Club. We asked them to tell us which of a long list of foods they had found difficult to resist when they were overweight. Twenty-five percent put cakes and biscuits at the top of the list, and a total of 72 percent named carbohydrate-rich foods as their main temptation. Sixty-four percent of the listed foods contained added refined sugar, while, of the other 16 foods mentioned, none scored more than 4 percent. These carbohydrate-rich foods, by the way, have another characteristic; they are all artificial foods that do not exist in nature in the form in which we eat them. As I have said elsewhere, people are not likely to get fat if they make up their diet mostly from the foods that were available to our prehistoric ancestors, like meat, fish, eggs, fruit, and vegetables, while as far as possible avoiding manufactured foods, most of which are carbohydrate-rich.

The fact is that, given the choice, people eat the foods that they like, and the more they like them, the more they are likely to eat them. You may think this so obvious that it is unnecessary to say it, but this simple fact accounts for most obesity. If you find that difficult to accept because of lack of proof, let me recall a story in Bernard Shaw's *The Adventures of the Black Girl in Her Search for God*.

In her wandering the girl comes across a scientist, clearly meant to be Pavlov, who is experimenting with a dog. When asked what he is doing, he says he has discovered that, when he shows the dog a piece of meat, the dog salivates. "But everybody knows that," says the black girl. "Maybe," answers the scientist. "But until I did the experiment, it wasn't scientifically established."

So what about establishing scientifically that the availability of very attractive foods causes obesity? During the last few years, research workers have discovered that the simplest way of producing a fat rat is not to offer it only the simple pellets that make up the very nutritious food normally given to rats, but to let it also have a go at eating cakes, biscuits, chocolates, and so on. Rats eat this sort of food with enthusiasm, and a very effective fattening diet it turns out to be. So it is now supported by experiment that such highly attractive foods promote overeating and obesity.

When you come to think of it, the fact that a low-carbohydrate diet is an effective way of losing excessive weight also suggests that obesity is caused by eating the irresistible high-carbohydrate foods. The low-carbohydrate diet severely limits just those foods that, as we saw, people find most tempting, while allowing you to eat as much as you like of foods such as meat and fish and vegetables. You lose weight because these last are foods the body needs to satisfy hunger, and not just to satisfy appetite, so you stop eating when you have had enough. This is not to say that these foods are in the slightest degree unappetizing; they do not, however, encourage overeating. It should also be remembered that low-carbohydrate foods are the ones that happen to contain a high concentration of the nutrients that the body needs.

Now let me try and explain why there are some people who consume quite a lot of sugar but are not overweight. There are three reasons why this can happen. The first would apply to those whose sugar intake is matched by a corresponding reduction in other foods, so that they are not taking excessive calories, although, as I shall show, they may be running the risk of nutritional deficiency. The second reason may be that they are extremely active people, so that they take a lot of calories but also use them up. The third possible reason why people might eat a lot of sugar and still not put on weight is controversial. There is now evidence that some lucky people's bodies have the facility of burning off surplus calories; sometimes this increase in metabolism is just the equivalent of the extra calories they take, and so they do not put on weight. This view is not universally accepted in the textbooks of physiology and nutrition, but I find the evidence is now quite convincing. Even these people, of course, have a limit to the number of surplus calories they can dispose of in this way; they too will put on weight if their intake of calories is in excess of disposal.

If you are one of the lucky ones who can get rid of excessive calories from sugar, you may not get fat, but by no means will you escape its other ill effects. Tooth decay, indigestion, diabetes, coronary thrombosis, and all the other conditions I shall discuss—these are not necessarily avoided by people who can eat lots of sugar without getting fat.

So there is no point in worrying whether or not everyone agrees that metabolism can increase in response to an increase of food consumption. Let us just say that you cannot help getting fat if you are taking in more calories than you can dispose of—and a very obvious and potent source of excessive calories is the consumption of foods and drinks that contain sugar, largely because people find them so delicious.

It may be that you are one of those who finds it difficult to accept that sugar can be an important factor in producing obesity. In America, especially, an intensive advertising and public relations campaign has been in progress for several years to convince the public that sugar has nothing to do with getting fat. First you are told that a spoon of sugar contains only 18 calories. The advertisements say: "Sugar's got what it takes. Only 18 calories to the teaspoon. And it's all ENERGY." This is quite true, provided you use a rather small spoon and make sure it is a level spoonful rather than the more usual heaped spoonful. Our research experience shows that most people take the sort of spoonful that gives them more like 30 calories than 18 calories.

You might want to work out how much sugar you take just in

tea and coffee. Suppose you take an average number of cups, which is about six a day. Suppose you take the not ridiculously large amount of two spoons a cup, each giving "only" 25 calories. That is 50 calories a cup, and 300 calories at the end of the day. This is what the whole truth is likely to be, rather than the partial and misleading truth about 18 calories a spoon.

There is also a second point. The sugar people tell you not only that sugar does not make you fat; they say it actually helps to make you slim. Their argument goes like this. People get hungry because they have a low level of glucose in the blood. If you eat sugar, you stop being hungry because it is very rapidly digested and absorbed, so that the level of glucose in the blood rises. Have a little sugar from time to time, then, and you will end up eating less, and so reduce your weight.

Here is a quote from a sugar industry advertisement:

Willpower fans, the search is over!
And guess where it's at? In sugar!
Sugar works faster than any other food to turn your appetite
down, turn energy up.
Spoil your appetite with sugar, and you could come up with
willpower.
Sugar—only 18 calories per teaspoon, and it's all energy.

Unfortunately, there are three flaws in this argument. The first is the idea that your eating is controlled by the level of your blood sugar. This theory has now largely been discarded. There is quite a lot of evidence that it is not correct, and certainly that it is not a complete explanation of what controls hunger. Second, there is no reason to believe that, just because it is absorbed quickly, sugar will affect your appetite any more than any other food will. Third, there is absolutely no evidence at all that the sugar reduces your hunger to an extent greater than the calories you have derived from it.

Suppose you have just taken two spoons of sugar in each of two cups of coffee, and thus gained 100 calories. You are now less hungry, so you eat less. But by how much? A hundred calories? Fifty calories? Three hundred calories? The only evidence I know of suggests that your appetite is reduced by *less* than the calories you have taken from the sugar. This evidence came from some tests I carried out some years ago, when the same "lose weight by eating more" story was being noised about, though in relation not to ordinary sugar (sucrose) but to glucose. The idea was that you took about one third of an ounce of glucose three times a day, a little while before each meal. Then you followed a calorie-restricted diet, and you were supposed to be able to do this more easily because the glucose had reduced your hunger.

What I did was to take two groups of overweight people and put them on the same calorie-reduced diet (one in fact designed by the manufacturers of the glucose tablets) with or without the additional glucose. At the end of six weeks, people taking the glucose had indeed lost weight, a matter of 6¾ pounds. But the people on the same diet without the glucose had lost about 11½ pounds— nearly 5 pounds more, or close to twice as much. You might think then that the glucose did nothing at all—that the people who consumed it ate the same amount of their diet as the others, but lost less because of the extra calories from the glucose. But in fact this would only account for about one pound of the difference, not the 5 pounds or so that we found. The only explanation seems to be that the glucose tablets actually *increased* the amount that people ate on their calorie-restricted diet—exactly the opposite of what it was supposed to do.

I suppose it is natural for the vast and powerful sugar interests to seek to protect themselves, since in the wealthier countries sugar makes a greater contribution to our diets, measured in calories, than does meat or bread or any other single commodity. But what is always sad is to see scientists being persuaded to support the sorts of claims I have just described. Is it because they like sugar just as much as other people do? Or is it because at least some of them have still not got around to accepting the idea that all carbohydrates don't behave in the same way in the body? Or is it that they have persuaded themselves that the modern scourge is too much fat in the diet and so they have difficulty in admitting that they may have been wrong?

Equally, it is difficult to see why any nutritionist should endorse the consumption of sugar at the present level. What with the high prevalence of obesity, there is no acceptable reason for recommending that sugar intake should *not* be reduced, or that it should be reduced only as part of a general reduction of food. It is after all the only food that supplies nothing whatever in the way of nutrients; it is, remember, the claim of the sugar refiners themselves that their product is virtually 100 percent pure. It supplies nothing whatever other than calories, and calories are all that matter in weight reduction.

Cutting down any other food—*any* other food—is bound to reduce nutrients as well as calories. There is no evidence that overweight people are taking an excess of nutrients; but there is quite a lot of evidence to suggest that some of them could do with a nutritionally better-balanced diet. I shall have more to say about this question of calories and nutrients in the next chapter.

The proof of the pudding is in the eating—or in this case in the not eating. Many people lose excessive weight very successfully simply by giving up sugar, or by severely restricting it. If you take only one spoon of sugar in each cup of tea or coffee, and you drink only five cups a day, you might lose ten pounds of weight in a year, just by eliminating the sugar in your coffee or tea.

Sometimes, in order to reduce their weight to acceptable levels, people also need to restrict starchy foods, and so adopt a strict low-carbohydrate diet. Of course, giving up sugary and starchy foods and sugary drinks requires some self-discipline, as does any alteration in dietary habits. But for several reasons, described in detail in my earlier book *This Slimming Business*, the low-carbohydrate diet is the most sensible and effective way of controlling bodyweight. And my colleagues and I have demonstrated by experiment, not simply by armchair calculation, that this kind of diet gives a far better supply of nutrients than is made available under the orthodox regime that involves eating the same foods as before, only less.

I have never really understood why so many doctors in the American medical and nutritional establishments have frowned upon a diet that tells you in effect to reduce only, or chiefly, those foods that give you the calories you *don't* need while giving you little of the nutrients you *do* need.

Although I said I was not going to go into details about the principles of obesity, I must add one important point about babies. I have already mentioned the custom, increasingly common among parents, of adding sugar to milk formulas and to the cereal and other foods on which babies are weaned, as well as giving them sugary drinks. The result is the number of fat babies to be seen everywhere, to the extent that pediatric authorities in the United States and the United Kingdom have frequently drawn attention to the problem.

A few years ago it was suggested that the overfeeding of babies not only made them fat, but encouraged the development and persistence of obesity when they became adults. The story was that the fat cells in the babies' adipose tissue are encouraged by overeating to divide, so that not only do the existing cells become full of fat but the body produces more cells in order to accommodate still more fat. This idea was based on the finding that the number of cells that could be seen in the adipose tissue of fat babies was greater than the number in thin babies. These extra cells persist into adulthood, so that the fat baby becomes an adult with more adipose tissue and hence a greater propensity to store fat. Such a person, it was concluded, would clearly have a greater problem in controlling excessive weight than one with a normal amount of adipose tissue.

More recently this suggestion has been challenged on the grounds that it depends on the ability to count accurately the number of cells in the adipose tissue. The critics say that in thin babies some of the cells are empty and can easily be missed when they are counted; in a fat baby the cells all contain fat and are therefore more visible and likely to be counted. This leads to what the critics believe is the mistaken conclusion that there are fewer fat cells in thin babies than in fat babies.

Whatever the truth about fat-cell numbers, what is certain is that babies, like older children and adults, get fat if they take in more calories than they use. And you have only to look around you to see how easy it is for a baby to get these excessive calories. Even though some manufacturers of baby foods have stopped putting sugar in their products, mothers will do so with little hesitation. Nor do they hesitate to give their babies sugary drinks in their bottles whenever they believe that the little ones are thirsty.

11

How to eat more calories without eating real food

A criticism that one hears frequently of refined sugar is that it supplies "empty calories." This is true. Often, the critics go on to say that the refining process is at fault in that it removes essential nutrients that are present in unrefined sugar in significant amounts. This is largely *not* true, as we have seen.

Having considered what happens when you take sugar in *addition* to your other foods, let us now look at what happens when you take it *instead* of some of your other foods. After all, if people take 500 calories a day as sugar, and sometimes much more, it is likely that there will be some reduction in other foods; there must be a limit to how much even the most gluttonous person can eat.

In the simplest situation, imagine a diet of 2,500 calories a day, made up largely of good nutritious foods like meat and cheese and milk and fish and fruit and vegetables, with some potatoes and bread and breakfast cereal. Now keep the calories at 2,500 but replace 500 or 550 of them by sugar, the average amount taken in a day. I have shown that you can usually do this simply by adding only moderate amounts of white sugar to your tea and coffee, and taking an occasional sugar-sweetened soft drink. Clearly, the result of this replacement of 20 percent of your calories by sugar would be a reduction in your intake of nutrients—protein, all vitamins, all mineral elements—also by 20 percent.

No nutritional deficiency will occur if your previous diet contained an excess of 20 percent of all the nutrients you required. But suppose it did not contain this surplus. More important, suppose that you were one of those who takes more than the average amount of sugar—equal perhaps to 30 percent of your calories, or even 40 percent. Now it begins to be more difficult, as you can see, to imagine that the diet of 2,500 calories which originally supplied as much of the nutrients as you needed will still do so when the foods containing them are replaced by 30 percent or 40 percent of nutrient-free food.

This does not mean that by eating 4½ or 5 ounces of sugar a day—or even 7 or 8 ounces—you would be rapidly heading for pellagra, beri-beri, or scurvy. In extreme cases, with quite a lot of sugar and with the remainder of the diet not too well constructed, such diseases do occasionally occur. I shall later refer to the role of sugar in producing full-fledged protein deficiency in poor countries. But it may very well occur that your diet is marginally insufficient in nutritional terms, so that you are in that twilight zone between excellent health and a manifest deficiency disease: not quite well; tired and easily exhausted; prone to aches and pains and odd infections. All these vague but very real symptoms occur in all of us at some time or another. But while being a bit under par is no proof that your diet is deficient, this must be considered as a possible cause in people whose diets are unbalanced by a large intake of sugar.

Is there any way of showing that sugar can really—not just hypothetically—push more desirable foods out of the diet? One way of finding this out, I thought, was to check the trends of consumption of different sorts of food, especially those that are universally recognized as highly nutritious—meat, milk, fish, fruit, and eggs. In particular I decided to look at the trends for meat, for two reasons. First, it falls into the category of highly nutritious foods, and second, for most people it is also highly palatable. I argued that the increase in the consumption of sugar-containing foods, because they too are very palatable, might be accompanied by a reduction in the consumption of meat.

I must break off to explain why, when you look at the relevant statistics, you have to bear in mind two important considerations. The first is that, although *total* sugar consumption in America stopped rising some 30 or 40 years ago, and in Britain in the last 12 to 15 years, there was a simultaneous decline in the use of sugar in the home and an increase in the amount of sugar used in manufactured food. Crudely, and not completely accurately, one can say that people are putting less sugar in beverages at home but take more sugar in ice cream and cakes and biscuits, where, incidentally, it comes with plenty of other calories but not much in the way of nutrients. You would then expect the effects of sugar in pushing other foods out of the diet to be increasing, even though the absolute amount of the sugar itself is not increasing.

The second point to bear in mind is that the foods I mentioned, besides being among the nutritionist's favorites, are also relatively expensive, so that more of them tended to be consumed by wealthy people than by poorer people. This social gradient has declined in the Western world with increasing affluence; the poorer sections of the population are not as poor as they used to be. So what nutritionists and economists have been predicting is that general increasing affluence would bring about an increasing consumption of meat, milk, fish, eggs, and fruit. One would expect little or no change in consumption by the wealthier groups of the population, who presumably were always able to eat as much of these desirable foods as they wished; on the other hand, one would expect a great rise in the amounts that poorer groups consume as their economic situation improves.

So what about my hunch that sugar and sugar-rich foods are driving these better foods out of our diets? We have been able to show that, in the USA, the gradual improvement in living standards has been accompanied by an increase in the consumption of fruit by the poorest section of the population, but at the same time by a significant decrease in the wealthiest section. In the UK, what we did was to look at the change in consumption of the nutritionally more desirable foods between 1936 and 1983, for both the poorest tenth of the British population and the wealthiest tenth. The undoubted improvement in the standard of living during the half-century was reflected in a significantly improved diet among the poorest section of the population. In the 1980s they were taking more than three times as much milk, twice as many eggs, nearly twice as much fish, and 50 percent more meat. But for the wealthier tenth of the population, the figures that we were able to collect for 1936 and 1983 showed a significant reduction in all of these items. The consumption of milk, meat, and eggs had fallen by about 30 percent, and of fish by more than 50 percent.

As for meat, everyone with any experience of the country before the Second World War knows that the poorer people ate little meat (see, for example, the famous studies of John Boyd Orr). Yet in spite of a sizable increase among the poorer people, average meat consumption in the UK has hardly changed since before the war. This can only have been due to a *decrease* in consumption by the wealthier people.

More recent evidence comes from the USA, where, as you probably know, there has been a considerable outcry by experts in the last few years about the existence of nutritional deficiencies. How much of a deficiency exists is uncertain. What *is* certain is that it is much more than most people had thought.

It is unlikely that the fall in the nutritional quality of the average American diet was due to increased economic hardship. The more likely explanation is again that some of the nutritionally good foods were being crowded out by the nutritionally inferior, sugar-based, foods. This is also the belief of Dr. Joan Courtless, a member of the U.S. Department of Agriculture, who says: "The surveys themselves show that it [the worsening of diets] lies in the choice being made—increased consumption of soft drinks and decreased consumption of milk; increased consumption of snacks and decreased consumption of vegetables and fruit." And "snacks" contain large amounts of sugar.

12

Can you prove it ?

If reading this book convinces you that sugar is in fact pure, white, and deadly, you will certainly get involved in a lot of arguments when you try to convince other people. It will help you and stop your being thrown off balance if you carefully consider not only the facts I shall be giving you, but the wider problem of how to weigh evidence about the causes of disease, and how to form your final judgments about these. Before I begin to talk about diabetes and heart disease and duodenal ulcers and several other conditions, I shall discuss this problem in general terms.

As you will see, quite a lot of the conclusions I shall be drawing in this book will inevitably be based partly on factual evidence and partly on personal judgment. Those of you who have attempted to follow reports on the enormous amount of research into the problem of heart disease that has been done and is continuing will not be surprised when I say that you have to mix objective facts and subjective opinion. Absolute proof of the cause or causes of any disease is sometimes not possible.

To get *absolute* proof that cigarette smoking causes lung cancer, you would need to take, say, 1,000 young people at the age of 15; pair them off as carefully as you could into two very similar groups of 500 each; make one group smoke from that time onward; and rigidly prevent the other group smoking. Then, after perhaps 30 or 40 years, you could begin to see whether a significantly larger number of people in the smoking group had developed lung cancer.

Since this sort of experiment is clearly out of the question on ethical as well as on practical grounds, it is necessary to examine evidence that is largely circumstantial, and to judge this against a background comprising, one hopes, rational, and plausible general biological principles. I have tried to do this here. I have tried to recognize the limitations of all the evidence that is available, and in interpreting it I have tried to stand back and assess it chiefly on the basis of whether it makes sense, whether it fits in with what can be discerned about the rules that govern living processes and living organisms.

It is logical then to spend a few minutes looking at both of these aspects: to ascertain the kinds of evidence one can hope to find about the causes of disease and the limitations of this evidence; and also to see if general laws can be detected that make sense in relation to the maintenance of health. Since I am talking in this book mostly about sugar, and since the most important disease I shall be talking about is heart disease, I shall refer briefly to sugar and heart disease, but the same principles apply to any cause and any disease.

I ought also to say just a little about the word "cause," because I am going to talk quite a lot about sugar being a "cause" of a number of diseases. In the first place, it is quite certain that none of the diseases I shall be talking about are *caused* by sugar in the same sort of way that heat *causes* ice to melt. People differ in their susceptibility to disease, so that even in identical conditions— supposing you could produce them—one man might have a heart attack and another might not. This susceptibility seems to a large extent to be inherited, so one may say that your chances of getting a coronary are less if your parents, grandparents, uncles, and aunts have mostly lived to a ripe old age without having the disease; the chances are greater if many members of your family have had it.

In addition to this genetic factor, environmental factors also play a role in coronary disease. Most people accept the proposition that several environmental factors are influential and that these include leading a sedentary sort of life and smoking cigarettes. What I am hoping to show is that eating a lot of sugar is another environmental factor (or cause) in producing heart disease. I do *not* propose to show that sugar is the one and only factor involved in producing this disease, or indeed, any disease.

One more word about causes. If an event A sets off another event B, and if without A, B would not occur, then you can call A the cause of B. But suppose I throw a lighted match into my wastepaper basket, and my study and then my house burn down. Was the cause the lighted match? Or the loose paper in my wastepaper basket? Or the fact that my house contained lots of books and an excessive amount of wood? If *any* one of these factors had been different, the house might not have burned down at all. Alternatively, there might have been a short circuit in the electrical supply to my desk lamp, so the house might have burned for a reason quite unrelated to a lighted match.

I could say that if I eat sugary foods I get holes in my teeth. Then presumably the sugary foods are the cause of dental decay. But I might not get dental decay, in spite of these foods, if I have a high genetic resistance to the disease; or if I brush my teeth immediately after eating these foods; or if I know how to keep my mouth free of the bacteria that actually attack the teeth after being stimulated to multiply and to become active by the sugar in food. Is sugar then the "cause" of tooth decay? Or is it the bacteria? Or the lack of resistance of my teeth?

So in what follows, I do not expect to show you that a high intake of sugar is the one and only cause of the diseases I mention. I do hope to persuade you, however, that, whatever your heredity, and however much you may persist in habits that are involved in producing one or other of these conditions, your chances of developing it would be significantly reduced if you reduced your sugar consumption.

Now what about the sorts of evidence that a particular cause produces a particular disease? Broadly, there are two chief types of evidence, *epidemiological* and *experimental.* By "epidemiological," I mean evidence that there is an association between the intensity of the supposed cause and the presence of the disease. Such evidence deals with these sorts of questions:

Is heart disease more common in populations that eat more sugar?

If there has been an increase in the number of people suffering from the disease, has there also been an increase in the consumption of sugar?

In any population, has more sugar been eaten by the people that actually have the disease than by those who do not have it?

"Experimental" evidence is produced when you attempt to answer these sorts of questions:

Does the feeding of sugar to animals in a laboratory lead to heart disease?

Does removal of sugar from the diet reduce the chances of animals or people getting heart disease?

You may also ask rather less direct questions, such as: short of producing the disease itself, does the feeding of sugar produce the sorts of changes you normally find associated with the disease?

As to general laws, it seems to me that one or two biological principles ought especially to be remembered in these days of very rapid changes in our environment. First, living organisms can often adapt to change if it is not too rapid, nor too profound. If, however, the change is very rapid and very profound the organisms may succumb. It may be that in a population some individuals will be more resistant and may survive even though the majority has died. If the change persists, a new population may ultimately arise from the survivors, in which all the individuals will be equipped with this higher resistance. It is likely that, for a fairly considerable alteration to occur in a population, something between 1,000 and 10,000 generations are needed. In human terms this would range anywhere from 30,000 to 300,000 years.

The second principle is less obvious, but I believe it is a logical corollary of the first. It is that, if there have been great changes in man's environment that have occurred in a much shorter time than 30,000 years, there are likely to be signs that man has not fully adapted, and this will probably show itself in the presence of disease of one sort or another.

I know people tend to resent this thought, but I am convinced that you will not find an exception to this rule. Think again of cigarette smoking, which in the UK has increased from an average of 1,100 cigarettes a year per adult in 1920 to more than 2,500 cigarettes a year in 1980. Think of the rapid decrease of physical activity: the use of labor-saving devices, the widespread use of the car, the television, and radio; all of these have made affluent man into an extraordinarily sedentary animal during the past 30 or 40 years. And few today would deny that cigarette smoking is a potent cause of lung cancer and that both cigarette smoking and sedentariness are important causes of heart disease.

I could go on and point to the indisputable fact that every single new drug that has been introduced has sooner or later been shown to produce unintended bad effects as well as the intended good effects—though let me hasten to add that this is of little consequence if the good effects are important and the bad effects unimportant.

If then there is reason to be concerned about a dietary cause of a widespread disease, one should look for some constituent of man's diet that has been introduced recently, or has increased considerably recently. And by "recently" I mean over a short period in evolutionary terms, say, 10,000 years. Conversely, a dietary constituent is unlikely to be the cause of a common disease if it has been a significant part of man's diet for a long time—say one million years or more. If there is a constituent that is new or that now forms a much larger part of our diet than previously, one should also ask what has brought this about.

It is these considerations that should be borne in mind when one considers the total evidence that involves sugar consumption in the production of diseases in man. These considerations are so important that it is necessary to look at each of them rather more closely in order to understand their uses and limitations. This is what I propose now to do—not in very great detail, but sufficiently for you to understand why much of what I write in this book has been the subject of argument and disagreement, and why I nevertheless believe that the total picture is fairly convincing.

First, then, *epidemiology*. The questions seemed reasonably straightforward. How much disease exists in different populations? How does it relate to sugar consumption? And so on. But in fact there are no easy answers. Take the question of the prevalence of disease. To begin with, there is no record anywhere of how many people suffer from a particular disease at any given time.

For example, no one knows how many people in America or Britain have a duodenal ulcer, or even what exactly the prevalence is of dental caries in these countries. The diagnosis of duodenal ulcer, or the measurement of the precise amount of dental decay, is not easy and not sufficiently precise for all physicians to agree in every instance. And even if you were to hazard a guess as to the prevalence of duodenal ulcer or dental caries by counting, for example the number of cases treated in hospitals, you could not possibly find out the statistics for a country that lacks well-organized medical and dental services—and this applies to more than two thirds of the countries in the world.

You might imagine the situation to be easier with those diseases that are often fatal, because you could then look at the record for mortality. But once more doctors do not always agree about a diagnosis of coronary thrombosis or particular sorts of cancer. The causes of death that different doctors report in similar cases may therefore differ. Doctors in different countries tend to have different standards, and statistics from the less well-developed countries are again often quite unreliable.

Epidemiological studies also require a knowledge of food consumption—in this particular instance, a knowledge of sugar consumption. Now it so happens that it is easier to find out how much sugar is currently being eaten in a country than how much is being eaten of any other food. In almost every country in the world, all sugar is produced in factories. Consequently production, export, and import are well recorded. But even so, this information may not be sufficient for present purposes. It does not tell you how sugar is distributed through the population, and this it is most important to know.

Let me explain. Imagine two countries with exactly the same average consumption of sugar—suppose an average of 60 grams a day (about two ounces). Suppose that in one country most people eat about 40 grams, and relatively few eat over 100 grams a day. In the second country, quite a number of people eat very little sugar indeed, but a large proportion eat over 100 grams a day. If it needs at least 100 grams of sugar a day to produce coronary thrombosis, then more people would clearly be affected in the second country, even though the average for both countries is the same. I shall have more to say on this point later.

There is also the question of how long the disease takes to develop. It seems that coronary thrombosis and also diabetes show themselves only many years after their onset. Ideally, then, one wants to know people's sugar consumption over perhaps 20, 30, or 40 years. It is clearly impossible to get this information. One can only hope that a careful measure of consumption today will in many instances give at least some idea of whether people take a lot of sugar, or a moderate amount, or little, and also whether they have maintained this habit for much of their lives.

These then are some of the limitations of epidemiological evidence. One cannot of course ignore such evidence; the questions to be answered here are too important to discard any possible clues as to the cause of diseases such as coronary thrombosis. But you should constantly bear in mind the limitations of this type of evidence. Especially you should not be surprised if it seems less than conclusive; you may have to be content if it simply gives an idea about a possible cause that can then be followed up by research in other directions.

Under the heading of epidemiology, I also include evolutionary findings. Here the chief limitations are the uncertainty of some of the records. While most anthropological authorities take the view that man has been a meat-eater for several millions of years, they do not have an exact picture of what he ate and especially how much he ate of each food. The presence of large numbers of animal bones near human remains make it certain that he ate some meat, but it can be argued that meat was only a small part of his diet, that he ate mostly vegetable foods, and that these were bound to leave far less in the way of evidence compared with animal bones. This is not the place to argue the matter in detail, but I am in agreement with the majority who hold that primitive man was largely carnivorous.

All in all, epidemiological evidence rarely provides conclusive proof of the relationship between diet and disease. It will, however, add important information to my case, and the total evidence will, I hope, be sufficient to convince you "beyond reasonable doubt."

So far, I have been talking about the epidemiological evidence that you get from a study of populations. But you can also get epidemiological evidence to do with the relation between a disease and its possible cause in individuals that make up a population. You can do this either after or before they develop the disease. For instance, you can find out whether people who have developed lung cancer were or were not cigarette-smokers: this is called a retrospective study. Or you can keep in touch with people of whom some are cigarette smokers and some not, and then see how many of each group later develop lung cancer: this is a prospective study. Whichever form your study takes, you will try and ensure that the people in the two groups differ, as far as you can discover, only in their smoking habits and are similar in every other way. You can see that it is easier to do this with individuals than with populations. If, for example, it is true that a population of rural Africans is much less likely to have cases of appendicitis than is a population of British town-dwellers, you could say it is because the Africans eat more fiber, or less fat, or less sugar—or indeed less food; or you could say that it is not because of their diet but because they do not ride in motorcars or watch television, but are physically more active. If, on the other hand, you find in the same town that men who have just had a heart attack have been eating more sugar than have men of the same age and social class who have not had a heart attack, you have a piece of evidence that sugar may be a possible cause of heart attack.

I now turn to *experimental* evidence of what causes disease, the sorts of ways in which this evidence is gathered and the ways in which it can be legitimately interpreted.

One of the best ways to understand human disease is to

reproduce the condition in rats or guinea pigs or other laboratory animals. By this means medicine has gained a good understanding, though by no means yet complete, of such hormone diseases as excess or deficiency of the thyroid hormone or the hormones of the pituitary gland or of the parathyroid glands. Again, most modern knowledge about nutrition—about calories and protein and vitamins and mineral elements—comes from work with laboratory animals.

On the other hand, when we can't produce a disease in animals, we are tremendously handicapped. There was a long delay before medicine found out how to treat pernicious anemia. This was because every suggested treatment had to be tried on patients with the disease. After very many years of hard research work, it was discovered that the eating of raw or lightly cooked liver was effective. Then, whenever a new extract from liver was made, it had to be tested in a patient with untreated pernicious anemia.

It was only after an interval of 23 years that this work ultimately resulted in the discovery that the active therapeutic agent in liver was vitamin B_{12}. There is no doubt that this long interval would have been very much reduced if the researchers could have been conducting the same tests on rats or rabbits or some other animal in the laboratory.

Coronary disease as it occurs in man has not been produced in any of the ordinary laboratory animals. There have been suggestions that it has been produced in some primates, but no one yet knows whether this can be done regularly and at will. In any event, it is going to be extremely difficult and costly to set up a laboratory with the hundreds of monkeys that would be necessary in order to carry out all the experiments needed to get somewhere near solving the problem of coronary disease.

What can be done more easily is to try and reproduce in laboratory animals as many as possible of the characteristics that are found to be commonly associated with the disease in man. The one characteristic that everyone has talked about for years is a raised level of blood cholesterol. It is widely accepted that the chances of someone developing a heart attack are higher when blood cholesterol is higher. It is reasonable, then, to suppose that the experimental manipulation of the diet or of other conditions that raise the level of blood cholesterol in animals may be concerned with producing coronary disease in man. As everybody knows, there have been thousands of these sorts of experiments. But people with coronary disease often have other abnormalities as well as an increase in blood cholesterol, and producing these experimentally can also help in identifying what causes the disease.

One important characteristic of coronary disease is the occurrence of those changes in the arteries known as "atherosclerosis," which are described in a later chapter. Not all animals are equally susceptible to this condition. It is relatively easy to produce changes in the arteries of rabbits, but much more difficult in rats. And when one does produce fatty changes in the arteries, there is always the question as to whether they are the same as those that occur in human atherosclerosis. There was for a long time—and there is still in some minds—a doubt whether what is produced in the rabbit is really similar to the condition in people. Occasionally, enthusiastic research workers are carried away sufficiently to claim that they have produced experimental atherosclerosis when what they have really produced is something demonstrably and grossly different.

What one would like to see would be experiments that produce many of the characteristics of coronary thrombosis all together in the same animals by the same means. Still better, if it is not possible to produce coronary thrombosis itself, would be to see the same experiments carried out in several species of animals, so no one could be misled by some unusual response in the particular species that happened to be studied.

One ought also to take into account something of the normal habits of the animals. If, for example, we are studying the effects of diet, it does not seem to me to be sensible to include foods that are not normally part of the animal's diet or normally part of a human diet. The diets of herbivorous animals like the rabbit ordinarily contain very little fat and virtually no cholesterol. It is not surprising to me that diets high in fat and containing cholesterol produce abnormalities in rabbits. I do not believe that this should be accepted as proof that similar diets will produce similar effects in carnivorous or omnivorous species of animals, including human beings, who have consumed such diets for hundreds of thousands of years.

As well as experimenting with animals, one could also do experiments with human beings provided one could be sure that no harmful effects would be produced. The intention would clearly be not to produce coronary thrombosis but to produce temporarily the sorts of changes that are known to accompany the disease. Once again, the commonest change that has been looked for is an increase in the level of cholesterol in the blood.

Let me break off for a moment to make a point about what sorts of measurements research workers carry out when they perform experiments such as those I have been talking about. Of course, a most important guiding principle is to measure the substances that, like cholesterol, are known to be considerably changed in concentration in the condition that is being examined. But research is often limited by the methods available for carrying out the measurements. It may be that for a particular substance either no method exists or none that is suitable for routine use, since it may require very special apparatus or may involve very laborious techniques. Conversely, it may be that quite simple methods exist for measuring the substance, even though it may not be—or may turn out not to be—very relevant to the disease being studied.

This, I believe, is the position with studies on coronary disease. If it is true—and I am still far from convinced—that the most important change in this condition is the increased level of some of the fatty materials in the blood, then there is a lot to be said for the view of many workers that levels of other fatty substances are more informative than levels of cholesterol. One such substance is triglyceride; another is one of the particular compounds that holds the cholesterol in the blood. This is the cholesterol bound to high-density lipoprotein ("HDL cholesterol") which is now accepted as a better indicator of coronary risk than the total amount of cholesterol. Indeed, not everyone is convinced that, in most people, much information is gained from measuring only total cholesterol. One distinguished American research physician has written that blood cholesterol is a biochemical measurement still in search of clinical significance!

One final sort of experimental evidence is to see what will cure or prevent the disease; from this, within reason, one can draw conclusions about the cause. An obvious example is scurvy, which is cured by giving oranges or lemons; it was this discovery that ultimately led to the identification of the cause of scurvy: a deficiency of vitamin C, which is contained in fruits and vegetables.

But there are two possible ways in which one can be misled, one obvious and one less obvious. The obvious one, though often overlooked, is that there are some conditions, like the rheumatic diseases, in which the symptoms fluctuate. A period of pain is frequently followed by a period of remission, so that any treatment given while the disease is worrying the patient is likely to be thought to have produced the subsequent improvement.

One can also fall into a rather more subtle trap. I can best explain this by an example. Many older people who suffer from a variety of diseases gradually develop a degree of heart failure, and one of the effects is swollen legs due to dropsy (edema). This can be relieved if large amounts of vitamin C are taken, for the vitamin acts as a diuretic and increases the loss of fluid through the kidneys. Yet although this treatment is curing a symptom of heart failure, the condition is clearly not due to a deficiency of vitamin C.

A more obvious example, if perhaps a rather ridiculous one, is that curing a headache with aspirin does not imply that the headache was caused by aspirin deficiency.

Let me here refer to the results of experiments on the effect of changing the diet in attempts to prevent coronary disease. Since these experiments have all been designed to test the effect of altering the fat content of the diet, and not the effect of altering the sugar content, it will be best to discuss these experiments at this point rather than later, when I shall be concentrating on the case against sugar. It is also useful to do this here because I shall be able to demonstrate another of the difficulties of research into the subject of diet and heart disease.

There have been several experiments, or trials, in which fat intake was changed by reducing the amount of saturated fats like butter fat and meat fat, sometimes also adding vegetable oil like corn oil. In some trials, the doctors studied people who had already had one or more attacks of coronary disease. The research workers tried to see whether the change of diet reduced the patients' chances of getting another attack compared with a similar group whose diet had not been changed.

This sort of study is called a "secondary prevention trial." The other sort of study is the "primary prevention trial," in which the investigators change the diet of apparently healthy men and see how many develop coronary disease, again compared with a matched group whose diet has not been changed.

Several trials of each sort have now taken place, the most important being those attempting primary prevention. One was in a veterans (ex-service) center in Los Angeles, in which 424 men were put on an experimental diet with reduced saturated fat and cholesterol, and increased polyunsaturated fat. During the next five years the researchers compared these men with 422 men whose diet was not changed. It turned out that there were 63 deaths from coronary disease in the experimental group and 82 in the control group; however, the number of deaths from all causes was the same in both groups. One unwelcome result was that more of the men in the experimental group developed gallstones.

A trial in Helsinki lasted 15 years. This involved patients in two different mental hospitals. In one, the patients received the standard Finnish diet, and in the other a diet with a high proportion of polyunsaturated fats. After five years the diets were switched. At the end of the 15-year experiment there were somewhat fewer coronary deaths in the two hospitals during the time patients were taking the experimental diet. However, in this trial, too, there was no difference in the total number of deaths from all causes. The way the experiment was conducted has been strongly criticized: for example, as patients left either of the hospitals, other patients were taken in to replenish the trial numbers, so that the study was conducted not with a constant population, but with patients who had been in the experiment for widely varying times.

Later trials have examined the effect of changing other items as well as the amount of dietary fat, although they have not changed the proportion of polyunsaturated fat. One such study was in the United States MR FIT (Multiple Risk Factor Intervention Trial), reported in 1982. More than 12,000 middle-aged men with high coronary risk factors took part, and half of them were given instructions aimed at reducing their cholesterol level, blood pressure, and smoking habits. They were frequently interviewed and encouraged to persist in these measures. After seven years their blood cholesterol concentration had fallen by only 2 percent, their mortality from coronary disease was not significantly different from that of the control group, and their total mortality was somewhat higher. The cost of this study was over 100 million dollars.

In 1984 there were reports of the Lipid Research Clinics Primary Prevention Trial. The purpose of this trial, however, was to study not the effect of diet, but the effects of the administration of a cholesterol-reducing drug, cholestyramine. Like the MR FIT experiment, the subjects were men—about 2,000 of them—selected because of their high coronary risk, in that they had blood cholesterol values in the highest 5 percent of the values in the population. All were instructed to take a low-fat diet and half of them were given the cholestyramine. At the end of seven years, both groups showed a decrease in their blood cholesterol, and the reduction was significantly more in the drug group. These also had nearly a quarter fewer deaths from coronary disease and a significantly smaller number of nonfatal heart attacks. Unfortunately, this desirable effect of the drug was accompanied by quite unpleasant gastric symptoms, so that many of the men in the drug group gave up taking it. Clearly, the mass treatment of the population with this drug is not really practicable. Moreover, it would be enormously expensive; it has been calculated that it would cost about a quarter of a million dollars to prevent a single heart attack.

A trial that is still proceeding is the Stanford Heart Disease Prevention Program. This began with a massive publicity campaign in two Californian towns, involving television and radio programs, regular newspaper articles, advertisements, and propaganda material through the post. The effect was a 3 percent reduction in the concentration of blood cholesterol, which after the end of the campaign reverted to only a 1 percent reduction. In spite of this, the next phase planned is a repetition of the trial on a bigger scale, involving five Californian towns.

We must conclude that all this effort since the 1960s has not succeeded in demonstrating the efficacy of a change in dietary fat in reducing the prevalence of coronary disease. And yet the fact remains that, since about 1960 in the USA, and somewhat later in other countries, there has been a decline in the number of people recorded as dying of coronary disease. In the USA, there has also been a small decrease in total fat consumption, but this dietary change began some years *after* the fall in coronary deaths began. There have also been other changes in lifestyle, both in America and in the UK, as well as in some other countries, although not all of them are quantifiable. These changes include a decrease in cigarette smoking, an increase in physical programs such as jogging and "aerobics," a more widespread attempt to control high blood pressure, and a vast number of coronary bypass operations on people with heart disease. In most of these countries there has also been a small but continuing decrease in sugar consumption. The fact remains though that we do not know why there has been this decline in coronary deaths, but it is of course very welcome.

It may turn out in the end that people would reduce their chances of getting heart attacks if they took large quantities of polyunsaturated fats such as those found in maize (corn) oil or in sunflowerseed oil, or in the special sorts of margarines made with these oils. I have to say however that I think it very unlikely that this will happen. I believe that the best diet for the human species is one made up as far as possible of the foods that were available in our hunting and food-gathering days. The oils rich in polyunsaturates have been available only because of recent advances in agriculture and in the even more recent elaborate industrial techniques of extracting and refining oils; the complex chemical processing of these and other oils to produce margarine removes this product even further from the sorts of foods available to humanity during millions of years of evolution.

13

Coronary thrombosis, the modern epidemic

No one today can be unaware of the tremendous concern about the large number of people dying from coronary heart disease. In America and Britain, they account for more than one fifth of all deaths. In these and other affluent countries, at least one out of three men over the age of 45 will die of heart disease. It is not surprising that in books, magazines, radio, and television programs much has been made of this problem over the past 25 years. But I find that there is still such a lot of misunderstanding about the nature of heart disease that I had better clear the air with definitions and descriptions before going on to consider the causes.

It may well be that you already have what you believe is a nice, simple picture of heart disease and how it comes about. If so, it probably goes like this: There is a fatty material in blood called cholesterol. As you grow older, the amount of cholesterol in the blood increases, especially if you have food that contains too much meat fat or butter fat. Because of the high level of cholesterol in the blood, some of it tends to become deposited on the inside of the walls of the arteries, including the coronary arteries. These supply blood to the thick muscle that makes up the wall of the heart, which pumps the blood round the body. The gradual narrowing of the coronary arteries by the deposited cholesterol reduces the blood supply to the heart, and you then get pain in the chest when you exercise—angina, or, more correctly, angina pectoris.

The cholesterol deposits also encourage blood clots to form, so that sooner or later one or other coronary artery, or one of their branches, becomes completely blocked. As a result, the blood supply to a larger or smaller part of the heart is cut off, and then you have your heart attack—pain, unconsciousness if the heart stops, death if it does not soon start beating again.

This view of a coronary attack is oversimplified; it is sufficiently misleading for me to ask you to bear with me while I go through the story again in more detail, and more in keeping with the real events. Especially, I want to differentiate between what medicine *knows* is happening, and what research is still uncertain about.

Like any other organ of the body, the heart can be affected by many different sorts of disease, so that strictly speaking it is as silly to speak of heart disease as it would be to speak of arm or leg disease. What people usually mean by heart disease is what is variously called coronary heart disease, or coronary thrombosis, or myocardial infarction, or ischemic heart disease. Even this statement, however, is rather misleading because these conditions are not quite the same. You will understand the situation better if you try to follow the disease process as it affects the heart—in so far, that is, as science understands it. I say this because in many ways no one is as yet clear about the development of the condition, or conditions.

In human terms, almost everybody knows what I am discussing. One common picture is that of an individual, more usually a man than a woman, and most commonly over the age of 60, who is often apparently quite healthy until he is stricken with a severe pain in the chest. He may fall unconscious and may not recover; or the pain may gradually diminish and he is put to bed. If he does recover from his first attack, he may have subsequent attacks after a shorter or a longer time, with again the chance that one of these will prove fatal. Sometimes the events are different. The picture then is of a person, again often apparently well, who dies so suddenly that he has virtually no time to complain of pain or of any other symptom.

The course of events leading to the established disease or diseases is unfortunately not at all clear. Indeed, whatever I now write, however carefully, will represent the views of many of the experts in this field, or even most of them, but there will always remain some who will disagree with some or all of the events as I outline them.

Let me begin by talking about the so-called "deposit" on the inside walls of the arteries. The deposit is called "atheroma," the condition is called "atheromatosis." The word "atheroma" is Greek for "porridge," and refers to the irregular patches of yellowish material found on the insides of the walls of the arteries; these patches are sometimes called "plaques." No one is quite sure what starts the process. Many believe that it starts with an aggregation of blood platelets on or in the wall of an artery. The platelets are tiny discrete bodies in very large numbers, floating in the blood together with the red and white blood corpuscles. When they stick together in this way they encourage the formation of tiny blood clots. Around these clots there is gradually built up a mass of fatty material that includes a fairly high proportion of cholesterol. In due course, these patches become fibrous, much as scars form on a cut on the skin. It is the combination of atheroma and fibrous scars that leads to this stage being known as atherosclerosis. Later still the plaques may degenerate and become chalky and hard.

Atherosclerosis can occur in arteries all over the body, although it is more likely to occur in some sites than in others. It probably starts at quite an early age, perhaps in the teens; according to some authorities, it starts even earlier. As it develops it may begin to interfere with the flow of blood so that exercise may give you a pain in the chest because of narrowing of the coronary vessels (angina), or pain in the legs because of narrowing of the arteries to the legs (peripheral vascular disease, also known as thromboangiitis obliterans, or Buerger's disease).

In peripheral vascular disease, an increase in the extent of atherosclerosis leads to pain in the legs after you have walked for a shorter or longer distance. If the condition is not treated, there comes a time when the blood supply to the extremities is so diminished that a toe may begin to die of gangrene, or the whole foot, or even part of the lower leg. Treatment may consist of drugs that widen the arteries, or of operative procedures to improve the circulation by stripping the arteries of their atheromatous material.

In the heart, the coronary arteries may become increasingly blocked, resulting in more and more severe angina brought on by less and less effort. A more complete blocking may also occur, with or without previous angina. It could be that the blockage is due to a blood clot; this occurs more readily in an artery with atheromatous patches, partly because of the slow rate of flow of the blood and partly because the normally smooth interior of the artery now contains rough atheromatous material. But a block may also occur because the narrow coronary artery just goes into a spasm or contraction long enough to cut off the blood supply and cause a heart attack.

The outcome depends on several things. One is the size of the portion of the heart that was supplied by the artery before it becomes blocked. A second factor is the particular portion that loses its blood supply, because some portions are very much more important in keeping the heart beating than others. Thirdly, the outcome depends on whether the relevant section of the heart has blood vessels coming to it from a different direction, which can rapidly expand and bring enough blood to it by this alternative route.

If the affected section of the heart is small or relatively unimportant, the heart will stop for only a short time or not at all. If a portion of the heart has permanently lost its blood supply, that portion may die. This is called myocardial infarction and can be seen years later in the heart where the dead tissue has become replaced with scar tissue.

It seems that something quite different occurs in sudden death. It is probably also associated with severe atherosclerosis of the coronary arteries, but what appears to happen in this instance is that the heart stops beating normally and goes into a sort of very rapid shivering, known as "ventricular fibrillation." This renders ineffective the heart's job of forcibly and regularly pumping blood round the body, and death ensues very rapidly indeed.

It is important to remember that it is possible to have quite extensive atherosclerosis without any symptoms at all. If so, it will be impossible to diagnose the condition unless some of the ather- oma has proceeded to the extent of becoming chalky so that it shows in an X-ray film. Most if not all adults in the well-off coun- tries live with at least a fair degree of atheroma but if they have no symptoms it is usually impossible to tell whether they do have ath- erosclerosis, and if so how much or where.

I hope you do not think that this has nothing to do with the subject of this book. One of my main reasons for taking up research in this field was that I became more and more uneasy about the prevalent simplistic view of how people get coronary disease—the idea that it is just a matter of cholesterol levels in the blood. This idea is now so firmly held by so very many people that they end up believing that anything that increases cholesterol in the blood is likely to cause coronary disease, that anything that reduces choles- terol helps to prevent the disease or even cure it, and that anything that does not invariably increase the cholesterol in the blood must have nothing whatever to do with the cause of heart disease.

I know I am biased, but this picture—in my view a rather naïve one—has hindered a proper understanding of the disease and its causes and so a proper understanding of its prevention.

In fact, people with coronary disease are afflicted with very much more extensive disturbances than just a rise in the level of cholesterol in the blood. For one thing, there is a rise in other fatty components in the blood, especially the triglycerides, sometimes called neutral fats; many people believe this rise occurs more fre- quently than does the rise in cholesterol. There is also a *fall* in the HDL cholesterol. Secondly, other biochemical changes take place including disturbance of the metabolism of glucose or blood sugar in the same direction as that found in diabetes. There is often a rise in the level of insulin and other hormones in the blood, and some- times a rise in uric acid. There are alterations in the activity of sev- eral enzymes. The behavior of the blood platelets is changed.

One could produce a list of at least twenty indicators that often register abnormally high, or abnormally low, in people that have severe atherosclerosis, and only one of these is the frequent though not at all universal rise in the level of cholesterol.

If you seek further evidence about the possible role of sugar or any other factor in producing heart disease in man, you should bear in mind the complexity of manifestations of the disease. This is particularly important in the sort of experiments my colleagues and I have conducted with laboratory animals. I shall talk about these in more detail in the next chapter.

The first proponent of the idea that fat could be a cause of coronary thrombosis, and since then its most vigorous defender, was Dr. Ancel Keys of Minneapolis. In 1953 he drew attention to the fact that there was a highly suggestive relationship between the intake of fat in six different countries and their death rate from coronary disease. This was certainly one of the most important contributions made to the study of heart disease. It has been responsible for an avalanche of reports by other research workers throughout the world; it has changed the diets of hundreds of thousands of people; and it has made huge sums of money for producers of foods that are incorporated into these special diets.

As a result, too, a very great deal is now known about the effect of different diets upon the processes of metabolism in the body, and especially upon the processes of fat metabolism. And yet there is a sizable minority of research workers, of whom I am one, who believe that coronary disease is not largely due to fat in the diet.

Let me start to argue the case by looking more closely at the epidemiological evidence of the relationship between diet and coronary disease. From the beginning, a few people were a little uneasy about Dr. Keys's evidence. Figures for coronary mortality and fat consumption existed for many more countries than the six referred to by Keys, and these other figures did not seem to fit into the beautiful straight-line relationship—the more fat, the more coronary disease—that was evident when only the six selected countries were considered.

Also evidence began to accumulate that not all fats were the same; some seemed to be good, some bad, some neutral. At first, this was strenuously denied by Dr. Keys, but by 1956 or so these differences were accepted by him as they were by all other workers. The "bad" fats were mostly animal fats such as those in meat and dairy products (saturated fats). The "good" fats were mostly vegetable oils (polyunsaturated fats). The "neutral" fats were neither good nor bad; an example is olive oil (mostly a monosaturated fat).

It seemed appropriate to look much more closely at the figures of mortality and fat consumption than had been done hitherto, and this I did in 1957. By putting down all the information available from international statistics, I found that there was a moderate but by no means excellent relationship between fat consumption and coronary mortality, which did not become closer even when one separated the fats into animal and vegetable. A better relationship turned out to exist between sugar consumption and coronary mortality in a variety of countries. The best relationship of all existed between the rise in the number of reported coronary deaths in the UK and the rise in the number of radio and television sets.

Making this last point serves two purposes, I think. The first and more superficial is to illustrate the possible dangers of finding an association between two events, and then saying that one event causes the other. It is unlikely, you would suppose, that your chances of becoming a coronary victim are increased just by the possession of a television. But, in the second place, when you look more closely, this suggestion is not so stupid after all.

The factors that have been implicated in causing coronary thrombosis include several that are associated with affluence—sedentariness, obesity, cigarette smoking, fat consumption, sugar consumption. On the one hand, therefore, the incidence of coronary thrombosis will be higher in those countries in which there is greater affluence as measured by any index such as cigarette or fat consumption, but also by the number of television sets or motorcars or telephones. On the other hand, many of these indices of affluence are likewise indices of sedentariness. People who have a TV are likely to be physically less active than those who do not. So it is not entirely silly to point to these relationships.

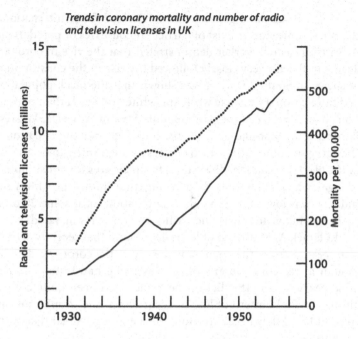

Trends in coronary mortality and number of radio
and television licenses in UK

The diagram shows the close association between mortality from
coronary thrombosis (right-hand curve) and the ownership of a
radio or television (left-hand curve). On the basis of this association alone, you might say that buying a radio or television will
increase your chances of having a heart attack.

Here I was, then, in 1957, with information from international
epidemiological studies suggesting that it would be at least as interesting to look at sugar consumption as to look at fat consumption.
There was no suggestion at that time that the existing studies were
a proof of the involvement of sugar. But, as in the story of fats, we
now had a clue. And, soon after my 1957 report, a Japanese research
worker confirmed the relationship between sugar intake and coronary heart disease in 20 countries.

Apart from these general figures derived from international statistics, some studies exist of particular countries or populations. A British research worker demonstrated that the rise in coronary deaths in Britain very closely followed the rise in the consumption of sugar. In South Africa, it was shown that the black population had little coronary disease while the white and the Indian populations had as much as the white populations in America, Western Europe, and Australasia. It seems, however, that the situation is changing in South Africa: heart disease is beginning to occur also in the black population. These facts fit the figures for consumption of sugar, which has been high for a long time among the whites and Indians, was low among the black population until some 20 or so years ago, but is now, with increasing affluence, rising rapidly.

In Israel, A. M. Cohen of Jerusalem found that recently arrived immigrants from the Yemen had very little coronary disease, though it was common among Yemenis who had immigrated 20 or so years earlier. The diet in the Yemen had been quite high in animal fat and butter but low in sugar; when the immigrants arrived in Israel they began to adopt the usual high-sugar diet of the country.

The Masai and the Samburu are two tribes in East Africa that live very largely on milk and meat, and thus have a very high consumption of animal fat. There is, however, very little heart disease among them. You might say that this is because they are physically very active. Another possibility is that they have a different sort of metabolism from that of other people, and recent work suggests that this is actually the case for the Masai. It seems that they have a more efficient way of dealing with animal fat without being subjected to a rise in the level of blood cholesterol. It is not clear, however, whether this is some genetic characteristic of the Masai or whether they have become so good at metabolizing fats because they have been coping with large quantities all their lives.

But what is often left out of these discussions is that both the Masai and the Samburu *eat virtually no sugar.*

Asian immigrants in Britain have a significantly higher mortality from coronary disease than do the native British—some 20 percent higher in men and nearly 30 percent higher in women. Yet a recent study has shown that the total intake of fat is almost the same in both communities, while the intake of saturated fat is lower and of polyunsaturated fat higher among the Asians. Thus, their ratio of polyunsaturated to saturated fat in the diet (the P:S ratio) is 0.85, compared with 0.28 for the native British. The high ratio in the Asian diet fulfills the recommendation of those who advocate changes in dietary fat in order to *prevent* coronary disease. It is clear then that the higher coronary mortality in Asians is not to be explained by differences in their fat intake. What was not measured in this study was the consumption of sugar by the Asians, but other investigations have shown that they in fact eat more sugar than do the rest of the British population. As we shall see later (p. 122), this is also relevant to the high prevalence of diabetes among Asians in Britain.

Let me quote only one other special study, made in St. Helena. Coronary disease is quite common on that island. This is not because the inhabitants eat a lot of fat; they eat less than the Americans or the British. It is not because they are physically inactive; St. Helena is extremely hilly and there is very little mechanical transport. It is not because they smoke a lot; cigarette consumption is much lower than it is in most Western countries. There is only one reasonable cause of the high incidence of coronary disease: the average sugar consumption in St. Helena is around 100 pounds per person a year.

In summary one can say that in most of the affluent populations I have considered, the prevalence of coronary disease is associated with the consumption of sugar. Since sugar consumption is, however, only one of a number of indices of wealth, the same sort of association exists with fat consumption, cigarette smoking, motorcar ownership, and so on. At this point it would be equally justifiable to look at any one of these factors as being a possible cause of coronary disease.

You can also put this rather differently by considering the relationship between any two of the factors I have mentioned. If you look at how much fat and sugar is eaten in different countries you find that they tend to be very similar for any one country; on the whole, both are low in poor countries, moderate in moderately well-off countries, and high in wealthy countries. So anything that is related to one is likely to be related to the other. You can now say, if you wish, that fat is a cause of coronary disease, and the association between sugar and the disease is accidental because fat and sugar are related. Or you can put it the other way around and say that sugar is a cause of coronary disease and it is the association with fat that is accidental.

When I arrived at this point it seemed to me that the next step was to look at the sugar consumption of individual people with and without coronary disease. For averages can be misleading; it is one thing to show that there is more coronary disease in countries where on average more sugar is eaten and quite another to show that, in any country, a person who eats more sugar stands a greater chance of getting the disease than a person who eats less sugar.

We devised what we thought would be a reasonably accurate way of getting at people's sugar intake, and measured this in 20 men with coronary disease, 25 with peripheral vascular disease, and 25 matched control patients (with other ailments) for comparison purposes. We spent a lot of time devising our method and choosing our subjects. The patients with coronary thrombosis, for example, were in the hospital with their first known attack, had up to this time no hint that they had heart disease, and had not consciously changed their diet.

We questioned them within the first three weeks after admission and asked about their normal diet before they were taken ill. We later showed that this method for measuring sugar intake was as good as the much more elaborate method normally used by nutritionists for other dietary constituents. We also showed that we were wise to have examined the diets of patients who had previously been apparently quite well. When we talked to them one or two years later, what they now called their normal sugar intake was in fact considerably lower than what they had reported on the first occasion.

In our study we found a very substantially higher sugar intake in the patients with coronary disease and with peripheral vascular disease than we found in the control subjects. The median values were 113 grams a day for the coronary patients, 128 grams for the patients with vascular disease, and 58 grams for the control patients.

When we published these results, there was a fair amount of criticism of both our conclusions and our method. We felt that much of this criticism was not valid, but in one regard there was justification. We had assessed sugar intake in our subjects by asking them in person, in the hospital, about their diets. Because of this personal contact we knew which was a patient with arterial disease and which was a control subject. It was possible that we were unconsciously biased by this knowledge and might therefore perhaps have exaggerated the sugar intake of the arterial patients and minimized that of the control subjects. In order to overcome this objection, we simplified our dietary questionnaire so that the patient himself could fill it in. The questionnaires were handed out by the ward sisters, and only after we had calculated the diets did we inquire about the category the respondents belonged to.

The results of our second study were similar to those of the first. The median sugar intake in the coronary patients was 147 grams; in the control subjects—this time there were two groups—it was 67 grams and 74 grams.

Since that time several other workers have examined the sugar intake of people with and without coronary disease. Some have confirmed our findings that coronary patients have been taking more sugar; some have not. There are, I think, several reasons for the negative results. First, people who have had a coronary attack are very likely to reduce their sugar intake, consciously or unconsciously, as we in fact found. You can just imagine what a shock it is to have had a "coronary" and how careful people will be to make sure they reduce their chances of getting another attack by keeping their weight down. The first thing people tend to do in this situation, is to cut down on sugar.

Second, we made certain that our controls suffered from no sort of condition that might affect their diet. So we chose healthy workers in a factory, or other patients who were in the hospital because of, say, a broken leg, but had no systemic condition. Third, we had found differences in sugar intake between different socioeconomic groups and between different age groups, so we made quite certain that our control subjects matched our arterial patients in these respects.

These are the sorts of reasons, I believe, why it is very possible that a less than careful selection of people to act as controls might lead to the false conclusion that there is little or no difference between the amount of sugar they eat and the amount eaten by people who develop coronary thrombosis.

It has, however, been said by my critics that, since not every investigator has found that individuals with coronary disease have been high sugar consumers, the sugar theory is entirely disproved. Most of these critics are, like Dr. Keys, strong supporters of the fat theory. The interesting point about this is that *no one* has ever shown any difference in *fat* consumption between people with and without coronary disease, but this has in no way deterred Dr. Keys and his followers.

Here let me deal with another criticism by the same people. They say that sugar cannot be a cause of heart disease because in the USA there was a considerable increase in that disease in the half-century up to about the middle 1970s, while sugar consumption hardly changed during that time.

But to make these criticisms is to misunderstand or misinterpret what you can reasonably expect from population studies. First, as I have often said, I believe that sugar is an important cause of heart disease, but certainly not the only cause. Sedentariness and smoking are only two of the other factors involved, and the incidence of both of these has changed a great deal during this century. Up till recently both had been increasing considerably, but people seem to have become more active during the past few years, and certainly many men have stopped smoking. Second, factors such as sugar and smoking and lack of physical activity take a long time to produce their effects, so that it is not easy to relate a time when changes occur to the time when they might affect the prevalence of coronary disease.

Third, it could well be that a high sugar consumption is more harmful in young people than in older people. We saw earlier that there has been a great increase in the consumption of soft drinks, ice cream, biscuits, and cakes; it is very largely young people that take these foods. The middle-aged have become increasingly figure-conscious and many have now reduced their sugar intake. So it seems likely that the constancy of the *average* consumption of sugar hides an increased consumption in young people and a decreased consumption in older people.

Lastly, and most importantly, it seems (as we shall see) that some 25 or 30 percent of people are sensitive to sugar, reacting to it in ways that could make them liable to heart attacks. If this is so, about three quarters of the population might be eating the same amount of sugar as the sensitive people are eating, or even more, but this would not make them suffer from heart attacks.

As I have said several times in this book, the epidemiological evidence cannot by itself *prove* that sugar or any other factor is a cause of coronary disease. It can only provide clues as to possible causes. We can then look for other kinds of evidence to see whether our theories hold water. Since I have so often been accused of saying that sugar is *the* cause of coronary disease, let me repeat what in fact I have said or written every time I have discussed the problem. Several factors are concerned in the production of coronary disease. One is genetic, that is, heredity; others are acquired. The genetic factor is responsible for some people being more susceptible to the environmental causes than others. Among the acquired causes are excess weight, cigarette smoking, physical inactivity— and also a high intake of sugar. It may turn out that they all ultimately have the same effect on metabolism and so produce coronary disease by the same mechanism. But this remains for further research to elucidate. In the meantime we must expect to find some people who get a heart attack although they don't eat much sugar, and some who have not had a heart attack although they eat lots of sugar; just as there are those who don't eat much confectionery but nevertheless get many holes in their teeth, and others who eat a great deal and have few holes in their teeth.

14

Eat sugar and see what happens

By the early 1960s I had decided that there was enough evidence from epidemiology to suggest that sugar might be one of the causes of coronary disease. The time had arrived, therefore, to begin to do some experiments to see what effects were produced by sugar in the diet. Since it seemed that the large increase in sugar consumption in Western countries was accompanied by a decrease in starch consumption to about the same extent, we fed our rats and some other animals in the laboratory with diets that contained all the protein, fat, carbohydrate, vitamins, and mineral salts that they needed, but varied the relative amount of starch and sugar in the carbohydrate part of their diets. Mostly the carbohydrate consisted either entirely of starch or entirely of sugar; sometimes it was a mixture of the two in a predetermined proportion. I should say that similar experiments were being carried out in other laboratories, notably by Professor Aharon Cohen of Jerusalem, who was however looking into the possible role of dietary sugar in producing diabetes rather than heart disease.

In our experiments with human volunteers, we asked them to record in detail the food and drink they took for a period of two weeks or more, with every item accurately weighed or measured, and written down at the time it was being consumed. They were then asked to increase the sugar they took—more sugar in tea and coffee and on their breakfast cereal, and more jam, confectionery, and other sugar items—while at the same time reducing the amount of starchy foods such as bread and potatoes. By the time they were due to change, we had calculated the amounts of all the elements of their ordinary diets during the preliminary period, and could now give them advice about how to make the change while maintaining the same total intake as before of carbohydrate, protein, fat, and other substances. This of course was not absolutely precise because we did not want to interfere with their normal lives more than we had to, but since they went on weighing and measuring their food, we knew whether and how much they deviated from the new diet. After two or three weeks on the high-sugar diet they went back to their usual diet while continuing to weigh and measure their intake for a further two or three weeks.

In our first laboratory experiment, we looked to see what sugar did in rats to the quantities of fatty substances such as cholesterol and triglyceride in the blood. We found that the amount of triglyceride in the blood was enormously and rapidly increased when rats ate sugar; the amount of cholesterol on the other hand did not change. Moreover, switching the diets resulted in very rapid change in the amount of triglyceride, which not only increased on the change from starch to sugar, but decreased again as sugar gave way to starch.

It later appeared, mostly through the research of other workers, that rats make and dispose of cholesterol quite differently from the way in which human beings deal with it. In other species, however, sugar was found to produce an increase in the amount of cholesterol, sometimes a considerable increase, as well as an increase in triglyceride. This occurs in baboons, chicks, pigs, and rabbits. In the spiny mouse, a desert animal, feeding with sugar produces such a considerable rise in cholesterol in the blood, and to a lesser extent in triglyceride, that these fatty materials give the blood a distinctly milky appearance. Moreover, while the liver of the rat becomes enlarged by some 25 percent, the liver of the spiny mouse increases to twice its normal size when the diet contains sugar.

In addition to the experiments on rats with normal diets, we have also used diets containing abnormal types of fats. By adding very saturated fats instead of the unsaturated fat that we usually use, and by adding a large amount of cholesterol to the diet too, we have produced much higher levels of cholesterol and of triglyceride. When we then substituted sugar for starch in these diets, there was a still greater rise in cholesterol and triglyceride.

Sugar produces many changes in rats beside the increases in cholesterol and triglyceride. I do not know how many and which of these will be found to be related to changes concerned with the development of atherosclerosis and coronary disease in humans. But I shall mention a few of the effects of sugar that at present seem to be linked to these conditions. I shall discuss still other changes later on in connection with other conditions in people.

Many research workers have studied the mechanisms by which the body makes and stores fat; the idea is that factors affecting these mechanisms may have something to do with the fatty materials that constitute atheroma. Along these lines, our studies included the measurement of some of the enzymes that are concerned in fat synthesis and storage. Our first measurements were of an enzyme in the liver called "pyruvate kinase." This enzyme is important in the production of fat in the body from a variety of substances derived from the diet. An increase in activity is taken as a measure of the fat-forming activity of the liver, the major site of fat synthesis. Young rats given sugar in the diet showed, after ten days, five times as much enzyme activity as did rats without sugar.

We also measured the activity of an enzyme complex called "fatty acid synthetase," which is closer to fat synthesis than is pyruvate kinase. It exists especially in the liver and in the fat tissue (adipose tissue). In the liver, an increased activity implies greater production of fat, which is then carried in the bloodstream. In adipose tissue, an increased activity implies a greater removal of fat from the blood for storage.

With a sugar diet instead of a starch diet for 30 days, rats developed twice as much synthetase activity in the liver, and one third as much in the adipose tissue. A rise in the liver and a fall in the adipose tissue suggests that more fat was put into the bloodstream by the liver. Nevertheless, there was no compensatory increase in the enzyme that would be responsible for storing this in the adipose tissue; there was, on the contrary, a decrease in this enzyme. We believe we have an explanation for this, to do with the fact that the hormone insulin is involved in converting the glucose part of sugar into fat, but is not involved in converting fructose into fat. This now gets into very complicated biochemistry, so I shall merely say that this is an example of the complex actions of sugar that I shall talk more about in Chapter 19.

The changes in enzyme activity that result from adding or subtracting sugar in the diet occur very quickly; in less than 24 hours you can detect the difference, and if you then change the diets over again, the process is reversed, once more in less than 24 hours.

I mentioned earlier that coronary disease in man is associated with a number of features other than the levels of fatty substances in the blood. So we looked for some of these features in our sugar-fed rats. The effects include an increase in blood pressure, a deterioration of the body's efficiency in dealing with high levels of blood glucose, a change in the properties of the blood platelets, and a change in the level of insulin in the blood. Rats fed high-sugar diets for a few months show all of these features.

Given a dose of glucose on an empty stomach, rats on a normal diet show a moderate rise in the blood level of glucose, which rapidly returns to fasting level. Rats kept on a high-sugar diet show a higher fasting level of blood glucose, a greater increase after the glucose dose, and a longer time before the level falls to fasting level. I shall have more to say about this behavior of glucose, "reduced glucose tolerance," when I discuss sugar and diabetes.

One cubic millimeter of blood contains about 250,000 of the small bodies called platelets, about 5½ million red blood corpuscles, and about 7,500 white blood corpuscles. If one cubic millimeter is an unfamiliar measurement, you can convert it into a more familiar unit by multiplying it by 1,000 to get the approximate numbers in one cubic centimeter. This figure multiplied by 5,000 gives approximate numbers in the whole body of an adult man.

The blood platelets are very much involved in the process of blood clotting. This is a highly complex process in which an important early step, or perhaps the very first step, is a change in the properties of the platelets; they become more sticky so that they can stick more readily to the walls of the arteries. They also clump together more readily.

These and other changes are common in people with severe atherosclerosis or coronary disease. We tested the platelets of our sugar-fed rats and found that they clumped together ("aggregated") distinctly more easily than did the platelets of the rats fed without sugar. The behavior of platelets is another matter that I shall bring up again later on.

I am increasingly inclined to believe that the clue to coronary diseases lies in a disturbance of the hormones of the body. This is why I think it important that Professor A. M. Cohen and others have shown that sugar-fed rats develop abnormalities in the way that the pancreas produces insulin. My colleagues and I have found in addition that sugar-fed rats also develop enlarged adrenal glands.

We have not been successful in producing atheroma in our rats because the strain of animals we use is resistant to the disease. But other workers have been able to do so. In Paris, Dr. L. Chevillard and his co-workers reported that rats develop atheroma of the main blood vessel, the aorta, when sugar is included in the diet.

Although atheroma did not develop in our rats, we analyzed the aorta to see if there was any difference in the fatty substances within the walls of this artery. We found substantially more cholesterol and triglyceride in the aortas of rats eating the sugar diet than in those eating the starch diet. We also looked at the effect of adding saturated fat or unsaturated fat to the diet, and found that it made no difference to the fatty substances in the aortic tissue.

I have been talking so far about our experiments with rats, since most of the experiments carried out by ourselves and by others on the effects of sucrose were done with these animals. However, some experiments with other animals have also been done. Rabbits fed sugar have been shown by us and by other research workers to develop a raised level of cholesterol. In cockerels and in pigs, we ourselves found that sugar increased the level of triglyceride. Our pigs also developed a high level of insulin in the blood. Cockerels of the Rhode Island/Light Sussex strain developed quite definite atheroma of the aorta with sugar, but not with starch. In a second experiment with White Leghorn cockerels, we measured the area of their aortas that was affected by fatty deposits. It came to 46 percent of the aortas in the chickens fed with sugar, and less than 1 percent in the chickens fed with no sugar.

What about human subjects? Professor Ian Macdonald of Guy's Hospital in London carried out many experiments with people who were given, mostly for a few days, mixtures of food components with and without sugar. Briefly, he found that, in young men, sugar raises the level of cholesterol in the blood, and especially raises the level of triglycerides. This does not happen with young women. It does happen in older women, however, after menopause.

Professor A. M. Cohen of Jerusalem has done experiments which for the most part were conducted over longer periods than those of Professor Macdonald, and his subjects were eating normal foods rather than mixtures of pure food items. They were given diets in which the carbohydrates were either mostly starch in the form of foods like bread, or mostly sugar. Professor Cohen and his co-workers found that the sugar diet produced a rise in cholesterol level, and also an impairment in glucose tolerance.

By now, it has been well established in several laboratories that sugar in the diet results in an increase in cholesterol and triglyceride in the blood of human subjects. Our own experiments have mostly involved the careful measurement of the usual diets of young men—then getting them to replace part of the starch with sugar while making as few other changes as possible. We carried out extensive examinations on these men while they were on their normal diet, again at the end of two or three weeks on the high-sugar diet, and then two weeks after they had gone back to their ordinary diet.

In our first experiments with nineteen young men, the sugar-rich diet produced an increase in blood triglyceride in all of them after two weeks. In addition, six of them showed other changes: they put on about five pounds in weight, the level of insulin in the blood rose, and there was an increase in the stickiness of the platelets. All of these changes disappeared entirely, or almost entirely, two weeks after the men went back to their usual diet.

Three aspects of these results we found especially interesting. The first was the fact that about a quarter or a third of our subjects showed this special sensitivity to sugar, while the remainder did not. This suggested to us the idea that only a proportion of men are susceptible to coronary thrombosis through eating sugar.

Secondly, the rise in the level of insulin recalled to us that there had been two or three British research workers who had suggested that a raised level of insulin could be a key factor in the production of atherosclerosis.

Thirdly, we were intrigued that the men who were susceptible to sugar, as shown by the rise in insulin, also put on a lot of weight while on sugar and lost in within two weeks of going back to their normal diet. This reminded us of the association between overweight and liability to coronary thrombosis. Indeed, it has been argued that, if eating sugar does increase the risk of heart attacks, this is only an indirect effect, since dietary sugar predisposes people to become overweight, and it is being overweight that predisposes to the disease. We tested this suggestion by getting some young men to overfeed by increasing either the sugar in their diet or the starch. With the sugar, there was an increase in the concentration of both triglyceride and cholesterol in the blood; with starch giving the same number of additional calories, there was no change in the concentration of either of these fatty substances.

Nevertheless, excess weight does increase the risk of developing heart disease. Moreover, many overweight people show some of the characteristics of the disease, including high blood pressure, increased glucose and insulin in the blood, and insensitivity of the tissues to the action of insulin.

One of the common features of people who are liable to have coronary disease is a raised blood pressure. Among the very few investigations that have been made to see if the blood pressure is raised when sugar is included in the diet was a study by Dr. Richard Ahrens from the United States, who worked in my laboratory for a year. He was able to demonstrate a small but definite increase of blood pressure in rats taking sugar. Later, he carried out a similar experiment with young men who were given diets containing varying amounts of sugar; they showed a rise in blood pressure proportional to the quantity of sugar in the diet. In reviewing the subject of sugar and heart disease, Dr. Ahrens wrote that the epidemic of coronary heart disease "continues to increase on a world-wide scale in rough proportion to the increase of sucrose consumption but not in proportion with saturated-fat intake."

Our suggestion that only some people get atherosclerosis from eating a lot of sugar led us also to suggest that there should be a difference between middle-aged men who have the disease and those who do not. People with the disease should include those who experience an increase in insulin from eating sugar, and there should therefore exist a relationship between the amount of sugar they eat and the level of insulin. Those who by middle age have no sign at all of atherosclerosis will include those who are not susceptible to sugar, so that there should be no relationship between their sugar intake and the level of insulin.

We tested this hypothesis on two groups, each consisting of 27 middle-aged men; one was a group of patients with peripheral vascular disease and the other a group of men with no symptoms who were coming to a clinic for a regular checkup. The results, plotted on a diagram, confirmed our prediction. On the whole, those patients who ate more sugar had higher insulin levels than did those who ate less sugar; among the "normal" people, those who ate more sugar had the same levels as those who ate less.

A second sugar-feeding experiment with 23 men produced several of the same results, but also some additional features. Once again, after two weeks on the high-sugar diet, all of the men showed a rise in triglyceride, and six of them a rise in insulin and platelet stickiness. This time, however, all the men also showed a distinct rise in blood cholesterol, and an *improvement* in glucose tolerance. I shall have more to say later about this effect on glucose tolerance.

Curiously enough, these additional results were not caused by a higher sugar intake in this experiment compared with the last; in fact, the average daily sugar intake was 300 grams compared with an intake of 440 grams in the first experiment. We believe that the fact that we do not always find a particular change when we give a high-sugar diet (for example, no increase in cholesterol levels in our first experiment but an increase in our second experiment) is due to the tremendous interaction of the changes produced by sugar and the ability of the body to counteract some of these changes by adaptation of its metabolic processes. This view will be further discussed later.

We asked those volunteers who had shown the rise in insulin and the other associated changes to help us with some additional experiments. In one of these experiments, we gave three of these men a high-sugar diet once more, and examined more closely the effects on the platelets; we also did the same with three of our volunteers in whom sugar had not produced a rise in insulin. We compared, that is, potential "hyperinsulin" people with "control" people. What we did this time was to look at the behavior of the platelets when they were suspended in blood plasma and subjected to a high electrical potential. This procedure, called electrophoresis, causes the platelets to move toward the positive pole at a particular speed. When a very small quantity of a substance called "adenosine diphosphate" (ADP) is added they move slightly faster; when one adds more ADP, the platelets move distinctly faster. At least, this is what happens with blood platelets from normal individuals. But platelet behavior differs among people with a variety of disease conditions, the most noticeable of which is atherosclerosis. Here, the platelets move much faster in the electric field with the low concentration of ADP, and more slowly again when the concentration of ADP is increased.

You will understand, then, that we were interested to see what a sugar diet does to the platelets both of people in whom it produces an increase in insulin and of people in whom it does not. We found the answer quite quickly. When they were taking their usual diets, the platelets of the three hyperinsulin men and of the three control men behaved normally; however, after ten days on the high-sugar diet, the platelets of the hyperinsulin men took on the behavior of people with atherosclerosis, while the platelets of the control people did not change. A week after the high-sugar diet, the behavior of the platelets of the hyperinsulin men began to revert to normal.

Another experiment with our hyperinsulin volunteers was conducted to see whether a hormone produced by the adrenal glands was affected as well as insulin. We asked eleven of them once more to go on a high-sugar diet. Before they did so, and two weeks after they had begun, we measured both insulin and a hormone from the adrenal gland related to cortisone. We found that the insulin level in fasting blood increased by about 40 percent after two weeks on the high-sugar diet; the level of the adrenal hormone, however, increased very much more, to between 300 and 400 percent of the original value. This observation recalls our finding that sugar produces an enlargement of the adrenal glands in rats.

We ended our research report by suggesting that these results could be used to screen people for their sensitivity to sucrose, or, as we said, to identify those people that were "sucrose sensitive." If a short period on a high-sugar diet produces a rise in insulin or adrenal hormone, we shall know that the subjects are in danger of developing coronary disease from eating too much sugar. If a high-sugar diet does not affect these hormones, then we shall know that sugar will not give them coronary disease, although of course it might still produce other ill effects. Unfortunately, we have been so busy with other research that we have not been able to pursue this idea. Nor has anyone else done so.

About six years after we published the results of these experiments, they were confirmed by Dr. Sheldon Reiser and his colleagues from the Nutrition Laboratory of the United States Department of Agriculture in Beltsville, near Washington, DC. For three weeks they gave women as well as men diets with either sugar or starch, switching the diets for the following three weeks. On the sugar diet, the men more than the women showed an increase in blood triglyceride, cholesterol, and glucose. But what was even more interesting to us was that the American workers confirmed our observation that a proportion of the subjects—a quarter or so—were especially sensitive to sugar, showing also an increase in the insulin concentration of the blood. In some of their experiments, they were able to show that "normal" quantities of sugar, about equal to the average American intake, was enough to produce these effects.

(Let me here insert a small anecdote. Some time after Dr. Reiser published the results of his research, I had a telephone call from an American medical journalist. He asked whether I had heard of Dr. Reiser's report, and if so did I think it was a breakthrough. I said that I thought it was indeed important, but that my opinion might be biased because Dr. Reiser's publication was a confirmation of our own work; however, because it was not new, it could hardly be called a "breakthrough." But didn't I agree, insisted the journalist, that it was at least an *American* breakthrough?)

Our view, then, is that the underlying cause of coronary disease is a disturbance of hormonal balance. Apart from increased insulin and adrenal hormone, for example, many patients show an increase in estrogen. We have only recently been measuring the concentration of this hormone in the blood of some of our volunteers. This was in some young men taking a diet in which they reduced their sugar intake from an average of about 150 grams a day to about 55 grams. After three weeks the concentration of estrogen fell from 11.5 units to 8.4 units; they then resumed their habitual diet, and after two weeks their estrogen concentration had risen again to 11.1 units.

15

Too much blood sugar—or too little

The way the body works is largely a matter of keeping the organs and tissues in a pretty constant environment inside the body. Anything, for example, that makes the level of sugar (glucose) in your blood fall below normal, or rise above normal, is promptly followed by actions that restore it to its original level. These actions are controlled partly by the nervous system but chiefly by the hormones. If for any reason the control mechanisms are not working properly, you will have an excessive amount of sugar in the blood, or a deficient amount, for part or all of the time. The condition of a high blood sugar is called hyperglycemia, and that of a low blood sugar, hypoglycemia.

Diabetes

The commonest cause of hyperglycemia is diabetes. Diabetes (more strictly, diabetes mellitus) is a disease that has been studied in very great detail for quite a long time—certainly over 100 years. Research workers are still, however, not at all clear about several features of the disease. In trying to summarize what we do know, I shall inevitably have to make it sound much simpler than it really is; I shall have also to be much more dogmatic than the limitations of our current knowledge warrant.

Broadly speaking, diabetes occurs mostly either in children or in middle-aged men and women. Juvenile diabetes tends to run in families rather more than does "maturity-onset diabetes." Again, when children with diabetes grow up, they are usually quite thin; maturity-onset diabetes is most commonly found in overweight people. Most patients with juvenile diabetes respond well to treatment with insulin, while most of those with maturity-onset diabetes are more resistant to the action of insulin. As a result, it is now more usual to classify patients as "insulin-dependent" or "non-insulin-dependent." Yet another way of classifying diabetes is into Type I and Type II. However, in practice it is quite common—especially among nonwhite (non-Caucasian) patients—to find individuals who do not clearly belong to either of these types. Soon after von Mering and Minkowski showed in 1890 that diabetes could be produced in the dog by the removal of its pancreas, it became evident that groups of cells in the pancreas called the islets of Langerhans were responsible for producing a substance that prevented diabetes. An effective preparation of this substance was made by Banting and Best in 1921. The substance was given the appropriate name "insulin" (*insula* is the Latin word for island).

It was natural, then, to imagine that all cases of diabetes were caused by a failure of the islets of Langerhans to produce enough insulin. But it is now known that this is not always true. On the whole, such a failure is the most common cause of Type I diabetes but not of Type II. The latter condition is often due to an insensitivity of the cells of the body to insulin. One of the most important actions of insulin is that of enabling the cells to utilize the glucose from the blood which is their main source of fuel. If, however, the cells have become insensitive to insulin, the pancreas produces more and more insulin in order to counteract the insensitivity.

It used to be usual to treat all forms of diabetes with injections of insulin. Nowadays, however, it is more common to treat Type II diabetes patients with drugs by mouth. These drugs mostly fall into two groups: those that increase insulin secretion by the pancreas, and those that seem to increase the sensitivity of the cells to the insulin that is already being secreted by the pancreas.

Even if their diabetes has been kept under quite good control, by insulin injections or by oral treatment, patients are likely after several years to develop a number of other conditions, including peripheral vascular disease and coronary thrombosis. In addition, diabetes can result in diseases of the eye—cataract and retinitis—and disease of the kidney. No one quite understands why these complications arise, although it may be partly because of long-standing abnormal blood-sugar levels, or because of other abnormal substances in the blood such as "ketone bodies." As I shall show in Chapter 19, there is reason to believe that arterial disease may arise from a continuing high level of insulin. I shall then discuss the interesting association between diabetes, overweight, and arterial disease, and the fact that people with any of these conditions are likely to have excessive insulin in the blood.

There are several reasons why I believe that eating too much sugar is one cause of diabetes—mostly of Type II diabetes, but possibly Type I too. First there is the epidemiological evidence. Much of it parallels what I have already cited for coronary thrombosis, but here the evidence is fraught with even more difficulties.

In some ways, one could have expected an association between diabetes and dietary sugar, or any other environmental factor, to be simpler than that for coronary thrombosis because diabetes is more readily diagnosed during life. But in fact not many countries have the facilities for the large-scale and fairly elaborate surveys that would be needed to detect early diabetes. And as for mortality statistics, the difficulty here is that people with diabetes often die of one or other of the many complications of the disease, and the death may then be certified as having been due to the complications rather than to the diabetes itself. So science is on rather uncertain ground about the prevalence of diabetes, and I can only give you the views that are commonly, but not universally, held by the experts.

They believe that diabetes in the well-off countries is much more prevalent today than it used to be. If you look for it carefully, by checking for sugar (glucose) in the urine, or testing the level of glucose in the blood, you can find at least mild diabetes in something like 2 percent of the population in Western countries.

Currently it is on the whole more prevalent in these countries than in the poorer countries. Among the people of Indian descent studied by Dr. G. D. Campbell in Natal, South Africa, there is a much higher prevalence than in India itself. The average intake of sugar in Natal is said to be 110 pounds or more a year; in India it is between 15 and 20 pounds a year. Moreover, there is much more disease among fairly wealthy Natal Indians than among the poorer.

One other epidemiological study worth mentioning is that of Dr. E. Ziegler of Switzerland. He compared the mortality due to diabetes in Switzerland with sugar intake, using a rather novel method of assessing this as the "sugar climate"—the total amount of sugar consumed over a period of years. He then demonstrated that the mortality from diabetes over a period of 20 years is correlated, both in men and in women, with this "sugar climate."

The view that diabetes may be caused by eating sugar has long been held by many people. The name "sugar diabetes" of course refers to the fact that sugar (glucose) is found in the urine of affected persons. But people also take the name to refer to dietary sugar as a cause of the disease as well as to one of its symptoms. Again, for more than 100 years before insulin was discovered, it was known that diets low in carbohydrates and especially in sugar were the best treatment for diabetes.

Yet the first detailed epidemiological evidence, put forward by Sir Harold Himsworth some 50 years ago, suggested that the disease was associated most closely with fat consumption. He showed that the mortality from the disease in different countries was often proportional to the average amounts of fat in local diets. But he himself expressed surprise that this was so, knowing that a diet high in fat was the currently accepted treatment for the disease. Himsworth wrote:

> The dietary factor which parallels these changes [in mortality and prevalence of diabetes] most closely is the consumption of fat, and this correlation is surprisingly consistent. . . . We are thus left with the paradox that, though the consumption of fat has no deleterious influence

on sugar tolerance, and fat diets actually reduce the susceptibility of animals to diabetogenic agents, the incidence
of human diabetes is correlated with the amount of fat
consumed.

Looking at the problem again some years later, I wondered
whether Himsworth's difficulty arose from making the common
assumption that all carbohydrate was equivalent. Since total carbohydrate consumption is similar in most countries, there was no
reason to suspect carbohydrates as a cause of diabetes. But when
you consider the different forms of carbohydrate, then you find that
the prevalence of diabetes is more closely related to the amount of
dietary sugar than to dietary fat. This is especially true if you take
into account the probability that it may take 20 years or so for the
diet to produce diabetes, as Dr. Campbell suggests.

When I related the number of people dying of diabetes in different countries to the amount of sugar or fat that was eaten some
20 years earlier, I found a high correlation with sugar and no correlation with fat. The sort of relationship with fat that is sometimes
found, and was found by Himsworth, comes about because, as I
pointed out, average fat consumption in different countries is frequently related to their sugar consumptions. The most likely explanation of the situation, then, is that sugar intake is a cause of
diabetes, and fat intake is only secondarily related to diabetes
through its association with sugar intake.

A year before I made these observations, a very interesting paper appeared from Professor Aharon Cohen in Israel. He examined
people for the presence of diabetes, and his study was especially
interesting for two reasons. First, it was made on Jews, who are said
to have more diabetes than non-Jews. Second, he was able to com-
pare people of four different backgrounds: people from Western
Europe and America; others from North Africa; others from the
Yemen who had recently arrived in Israel; and some from the Yemen who had arrived 20 or more years earlier.

All but the recent immigrants from the Yemen had a similar
prevalence of diabetes. But the recent Yemeni immigrants had a

prevalence of 0.06 percent compared with 2.9 percent for the earlier Yemeni immigrants. Later, Cohen and his colleagues showed, as I mentioned in relation to his study on heart disease, that the major change in the diet of the Yemenis in Israel was a great increase in sugar consumption; there was very little change in their fat intake.

While I was in the process of revising this section of *Pure, White, and Deadly* for the current edition, a paper appeared in the *British Medical Journal* reporting a survey of the prevalence of diabetes in 34,000 Asians and 27,000 Europeans living near London. It transpired that diabetes was nearly four times as common in the Asians as in the Europeans. According to Dr. Tom Sanders, who is working in the Nutrition Department of Queen Elizabeth College and has been making a special study of the diets of Asian immigrants, they eat significantly more sugar than do the Europeans among whom they live.

In addition to these epidemiological studies, there is now quite a lot of experimental evidence that sugar may produce diabetes. Again, some of the early studies were those of Professor Cohen, and my colleagues and I have confirmed his results. Rats fed with sugar develop a decreased glucose tolerance resembling the condition seen in diabetes. That is, when a dose of glucose is given by mouth to a fasted animal, the already high level of glucose increases to a still more abnormal level and does not return to the fasting level within the usual one and a half to two hours.

Cohen showed that this impairment of glucose tolerance occurred in rats after three weeks or so when there was 67 percent sugar in the diet, after six weeks when it contained 40 percent of sugar, and after about 13 weeks with 33 percent sugar. The glucose tolerance recovered after a few days on the normal diet. When sugar feeding was resumed it deteriorated again, but this time after only a few days.

Later, Professor Cohen worked for a few months in my department, and again we studied the effects of feeding sugar to rats. This time we injected tolbutamide, one of the drugs used in the treatment of diabetes. This stimulates the pancreas to secrete insulin, which lowers the blood glucose level. We argued that if the sugar diet had made the rat somewhat diabetic it would not be using glucose as well as it normally did; tolbutamide would then have a lesser effect in lowering the blood glucose.

This is just what we found. In one experiment, after eight weeks, the injection lowered the blood glucose by 31 percent in the starch-fed rats and by 26 percent in the sugar-fed rats. In a second experiment the figures were 32 percent and 27 percent.

In human subjects, a high-sugar diet maintained for several weeks had been shown to reduce sugar tolerance, and a low-sugar diet for several weeks has been shown to improve it. We ourselves measured glucose tolerance in the experiment with young men that I described earlier, in which they were fed a high-sucrose diet for two weeks.

In the first of these experiments we found no change. In the second experiment we found an improvement in glucose tolerance after one week, and a slight reversion toward the normal after the second week. This may seem strange; in fact it is not at all surprising. The first effect of the sugar would be to improve the body's use of glucose by the common process of adaptation. It would do this either by improving production of insulin from the pancreas or by improving sensitivity of the body tissues to the action of insulin. But by continuing to give a high-sugar diet, adaptation would diminish and exhaustion take its place, and the use of glucose would now be less than normal. Thus the improvement in glucose tolerance that we showed after one week would not contradict the deterioration that people found after several weeks. Nor would there be a conflict in the fact that we found no change in our first experiment; we might very well have made our measurements at a point where developing deterioration just about canceled out the initial improvements induced by the sugar.

Apart from the decreased glucose tolerance that is found in diabetes, there are other noteworthy characteristics of the disease. At this point it will be convenient to discuss these in some detail in relation to the experiments that we and others have done with sugar.

Long-standing diabetes often causes deterioration of vision because of the development of abnormalities in the retina, a condition known as "diabetic retinopathy" or "retinitis." Several years ago, Professor Aharon Cohen showed that dietary sugar produced abnormalities of the eye in the rat. By using a very delicate technique that measures the electrical response of the retina to a flash of light, he and his colleagues found a diminished response in rats fed sugar. This observation was followed by a more detailed study by a London group that included one of my colleagues; they concluded from careful biochemical and microscopic examination that the retinal abnormalities produced by the sugar were identical with those found in diabetic rats.

As well as producing an increase in the size of the liver, sugar in the diet also results in enlargement of the kidneys. Quite early on in the story of the research into the effects of dietary sugar, Professor Aharon Cohen showed that the kidneys of his sugar-fed rats were abnormal, with, among other things, an increase of fibrous tissue between the blood capillaries. After this discovery, we ourselves became increasingly interested in the effects of sugar on the kidneys. There were two reasons for this. The main one was our increasing realization of their close similarity to the effects of diabetes, and the second was the happy coincidence that Dr. R. G. Price of the Biochemistry Department of Queen Elizabeth College had for a long time been carrying out research on the biochemical changes occurring in various diseases of the kidney. He and his associates had found that a very early sign of damage to the kidney was the appearance in the urine of a considerably increased quantity of a particular enzyme. This has the rather elaborate name (even when shortened) of N-acetyl-ß-glucosaminidase, but is known familiarly as NAG. This is just as well, since its full name is in fact 56 characters long, as against the 23 of the "short" form.

Given a diet with sugar, rats show an increase in NAG in the urine, and so do human volunteers who increase their sugar intake. After the rats had been taking the sugar diet for a year, it was possible to detect small calcified deposits in the kidney. I would not claim that this proves that sugar can be one of the causes of kidney stones; if it is, it is certainly not the only one, since kidney stones occur in populations that take little sugar, and are known to have been common long before sugar became a sizable item of our diet in the wealthier countries. On the other hand, since most kidney stones contain calcium oxalate or uric acid it is perhaps relevant that dietary sugar has been reported as increasing the amounts of these materials in the urine. The researchers who did this work have also said that patients with kidney stones have a low glucose tolerance, like that found in diabetics.

Our own work on the kidney at Queen Elizabeth College, however, has revealed what I think is the most striking evidence of the relationship between dietary sugar and the development of diabetes. We examined the kidneys of sugar-fed animals with the electron microscope, which takes photographs at magnifications of 10,000 or more. We looked especially at membranes of the cells that make up the vast number of tiny filter units, the glomerular capillaries, where the blood is filtered as the first stage in the elaborate process of producing urine. We noticed that these cell membranes were much thicker than they normally are. This was especially interesting because thickening of what is called the "glomerular basement membrane" (GBM) is accepted as the most characteristic abnormality found in diabetes among patients who develop "diabetic nephropathy"—that is, kidney disease.

Proceeding from this, some very sophisticated biochemical procedures were carried out, in which the glomerular basement membranes were separated and measurements made of their constituents. We got good evidence for an increased production of GBM by showing that several of the particular chemical units making up the membrane were present in larger amounts in the sugar-fed rats, and that there was greater activity of the enzyme involved in making the GBM with these units.

These abnormalities produced by sugar are exactly similar to those present in rats that develop diabetes for other reasons.

The importance of this research can be judged from the fact that in the UK something like 15 percent of patients with kidney failure, whether or not they are being treated with dialysis or kidney transplant, have developed their condition from diabetes, while in America it accounts for 25 percent of patients undergoing treatment for kidney disease.

In Type II diabetes the main feature of the disease is not a failure of the pancreas to produce its normal quantity of insulin, but a failure of the body's tissues to react sufficiently to the insulin that is produced. This can quite easily be shown in the laboratory. A small piece of tissue is put into a vessel, and one or other of the metabolic processes involving insulin is measured. For instance, you can put some glucose with a piece of muscle tissue in a closed vessel and see how rapidly it uses oxygen, or produces carbon dioxide, as the tissue oxidizes the glucose. Or you can put a piece of fatty tissue into a vessel and measure the rate at which new fat is produced. If you now do the same experiment and add insulin, you find that the oxidation or the fat formation has been measurably speeded up. But if you repeat all this with a piece of tissue from a diabetic animal, or a person with Type II diabetes, the addition of insulin makes little or no difference to the speed of these reactions; in other words, the diabetic tissue is insulin resistant. (The same phenomenon, though usually less pronounced, can be seen in people who are significantly overweight; this is another fact that I shall return to later.)

Similar experiments have been carried out with animals fed sugar for some weeks. They were first reported by research workers in Czechoslovakia, and later by ourselves at Queen Elizabeth College. Both in muscle and in fatty tissue, the inclusion of sugar in the animals' diet produces insulin resistance. In one of our experiments, the rate of fat synthesis in fatty tissue from starch-fed animals increased by about 140 percent when insulin was added; in fatty tissue taken from sugar-fed animals, on the other hand, insulin produced no increase at all.

The difference between the effects of short-term feeding of sugar and the effects of longer-term feeding can be important, although they are often ignored. It is commonly said that the concentration of glucose in the blood before breakfast—the so-called fasting blood-glucose concentration, which is elevated in diabetics—is not affected by adding sugar to a meal on the previous day. Nor does this affect the glucose tolerance—the response of blood glucose concentration to a dose of glucose—nor the simultaneous response of the insulin concentration. But this must not be taken to mean that sugar may be consumed by a diabetic, even in moderate quantities, as a regular part of the diet. As we have seen, it only requires the regular consumption of sugar each day for two or three weeks to produce a significant decrease in glucose tolerance, and in susceptible people a significant increase in the insulin concentration in fasting blood.

Unfortunately, a lot of the recent research that claims that a diabetic can take sugar with impunity depends on the results of tests with sugar given in a single meal.

Finally, I should mention the relationship between diabetes and coronary disease, which works both ways. On the one hand, if you are a diabetic you have a greater than normal chance of suffering from coronary disease. On the other hand, if you have coronary disease you have a greater than normal chance of developing diabetes—or at least of having an impaired glucose tolerance that is sometimes called "preclinical diabetes." I believe this sort of overlap is important when you come to try to understand how sugar can be involved in causing these two diseases.

Hypoglycemia

The people who know this condition best are diabetics. Sooner or later they run into the situation of having taken too much insulin, or too much of one of the new oral drugs, and they get the very uncomfortable symptoms of hypoglycemia (a low blood glucose level), sometimes even leading to unconsciousness. But hypoglycemia also occurs in many people who are not diabetics, although they rarely get it so severely as to become unconscious.

You begin by feeling hungry and weak, and you may begin to sweat. You may then start shaking, feel faint, and dizzy, and get a severe headache. If the condition persists, you may get mentally confused, stagger about and speak indistinctly or nonsensically.

At this point you could even be arrested for being drunk and disorderly.

All these symptoms have arisen because your blood glucose has fallen to an abnormally low level. It is easy to understand how this happens to diabetics who may have taken their insulin or a pill to lower the blood sugar, and then missed their normal breakfast because of some interruption. It is also easy to understand how it occurs in the rare circumstances when a patient has a tumor of the pancreas causing an overgrowth of its insulin-making cells.

The way it happens in other people is most commonly because of the consumption of a lot of carbohydrates, especially sugar. The effect of eating any meal is to increase the level of blood sugar. If sugar or starch or glucose is in the meal, then all or part of it turns up in the blood quite quickly as glucose. If protein or fat is in the meal then their digestion products too will in part be converted into glucose, but more slowly; in addition, they slow down the absorption of all food.

The rise in blood glucose is only temporary, because one of its effects is to stimulate the pancreas to produce more insulin. This causes both an increase in the breakdown of the blood glucose and an increase in its conversion into glycogen to be stored in the muscles and liver. As a result, the level of glucose falls back toward normal. A more than normally rapid absorption of a great deal of glucose occurs if a lot of sugar is consumed, especially if it is taken between meals when there are no other food constituents in the stomach that might delay absorption. There is then a rapid rise of blood glucose, and an excessive amount of insulin is secreted. Because of this, the subsequent fall of blood glucose is excessive, the level becomes abnormally low and if it is low enough symptoms of hypoglycemia will appear.

There is some evidence, too, that continued high intake of sugar can, at least for a time, result in an increased sensitivity of the pancreas, so that it responds more readily still with increased secretion of insulin, and hypoglycemia becomes even more likely.

How then do you treat hypoglycemia? Well, if you don't bother to think out the consequences of the process I have just described, clearly you treat a person with low blood sugar by giving them a lump of sugar to eat, or a sugary drink. And the effect is pretty miraculous: within a few minutes all the sweating and weakness and dizziness disappear. But now think back for a moment and you will see that this, however effective, is in the long run just what should not be done, because the rapid rise in blood glucose may be followed by a rapid fall.

What you must do is to prevent these large swings in blood glucose. Only foods that result in a gentle rise in blood sugar should be eaten, so that an excessive output of insulin by the pancreas is not evoked. That is why the best treatment for a lack of sugar (glucose) in the blood is the paradoxical treatment of avoiding sugar (sucrose) in your diet as much as possible.

Let me say a word here about hypoglycemia in babies. Premature babies sometimes suffer from hypoglycemia, presumably because their hormonal control of the level of blood glucose has not yet become properly balanced. This can be quite serious, and premature babies have been known to become unconscious or even die from hypoglycemia. Because this is an acute and hazardous situation, the best treatment in such an emergency is to give them sugar (sucrose) or, still better, to give them glucose by mouth or even intravenously.

One would expect that babies not born prematurely would not develop hypoglycemia so readily but might still be rather more sensitive to the damaging effect of sugar than adults. When you consider how soon babies are given sugar, and how much, it is perhaps not so surprising that there appears to be an increase in the number of babies who develop hypoglycemia when they are a few months old.

There seems to be a belief, especially in America, that hypoglycemia is quite common. My own view is that, although hypoglycemia is not exactly rare, it does not occur as commonly as is often claimed. In particular, the repeated assertion (again especially in America) that dietary sugar may cause hyperactivity in children and delinquency in young people has not been substantiated. It is said that both of these conditions are related to hypoglycemia and can be cured by eliminating sugar from the diet, or at least considerably reducing it. In spite of the suggestion that these claims have been demonstrated by carefully conducted experiments, closer scrutiny of the methods used shows that the case if far from being proved.

The relationship between coronary heart disease and diabetes

I have described in some detail why I think sugar is one of the causes of diabetes, and also of coronary thrombosis. These are not the only conditions in which I believe sugar is involved, but they are probably the most important ones. Before I turn to these other conditions, however, I am going to summarize the arguments I used in relation to coronary disease, because—apart from bringing together what I have had to spread over many pages—it will also help to make clear the close relationship between coronary disease and diabetes.

We can best do this by outlining the major features of coronary heart disease. These are:

1. The wide range of abnormalities found in patients.
2. The multiplicity of causes, which include cigarette smoking, lack of physical activity, excess weight, peripheral vascular disease and diabetes.
3. The difference in incidence between men and women.
4. The association with other diseases, notably diabetes but also high blood pressure, gout, gall bladder disease, peptic ulcer, and peripheral vascular disease.

I have set out in the table some of the more important abnormalities found in coronary heart disease; all of these are also found in maturity-onset diabetes.

Features in which abnormalities are commonly seen in coronary heart disease and diabetes Type II

Coronary heart disease	In blood	In other items
	Cholesterol	Glucose tolerance
	Triglyceride	Insulin sensitivity
	HDL Cholesterol	Platelet aggregation
	Uric acid	Platelet electrophoresis
	Glucose	Blood pressure
	Insulin	
	Cortisol	
	Estrogen	
Diabetes Type II	as above, together with retinitis and nephropathy	

All of these abnormalities may be produced by dietary sugar.

It is difficult to believe that this wide range of abnormalities seen in heart disease can arise simply from a disturbance in the way the body deals with dietary fat, or simply from a disturbance in the body's control of the amount of cholesterol in the blood. It is much more likely that such a complex of relationships and abnormalities is caused by a disturbance of hormone balance. In particular, insulin, cortisol, and estrogen affect many of the body's functions and much of the body's chemistry. More than this, a disturbance in the activity of one of these hormones usually leads to a disturbance in the activity of one or more of the other hormones. It is then not difficult to imagine that the result might well be the laying of the foundations of more than one disease.

The suggestion that coronary heart disease is brought about by a disturbed balance of the body's hormones is not new, although some of the earlier suggestions have now been almost forgotten. The possible role of hormones may be inferred almost automatically from the considerable protection women have before the menopause. The original suggestion about hormone involvement was made as long ago as 1956. A group of workers then pointed out that young women with diabetes are especially liable to develop coronary heart disease, and suggested that their "loss of immunity to coronary atherosclerosis" could be due to the effects of the insulin injections they are given. And in 1961 another group of researchers wrote, "Clearly any statement regarding the etiology [cause] of coronary heart disease will have to explain the sex ratio," and they go on to say that this strongly suggests a hormonal cause of the disease.

Other workers too have suggested that coronary thrombosis could be due to an abnormally high concentration of insulin in the circulating blood. There are several pieces of evidence to support this suggestion, the most obvious being that most patients with the disease have a high level of insulin in the blood. In addition, several of the agreed causes of coronary disease are often accompanied by a high insulin concentration in the blood; these include cigarette smoking, excess weight, peripheral vascular disease, and diabetes Type II. Thirdly, loss of excess weight and increased physical activity, both of which reduce the risk of developing coronary disease, result in a fall in insulin levels. Fourthly, experiments with rats have shown that administration of insulin produces an increased amount of cholesterol deposited in the body's main artery, the aorta.

As for sugar, the most relevant fact is that every one of the abnormalities seen in coronary heart disease and in diabetes can be produced by the inclusion of sugar in the diet.

16

A pain in the middle

It was almost by accident that I became interested in the relationship between sugar and severe indigestion or dyspepsia. I had been involved in the study of obesity and in its treatment for a long time. For a number of theoretical reasons, I began several years ago to treat people with diets restricted in carbohydrate. At first, these diets were restricted *mostly* in carbohydrate, but also somewhat restricted in fat. After two or three years, however, I realized that it was necessary deliberately to restrict only carbohydrate, because it turned out in practice that if you do this, you automatically restrict your fat.

For several years I have recommended this diet to all the very many overweight people I have seen, in the hospital or in my university department. As I point out in Chapter 2, such a diet more closely resembles what our ancestors ate during at least two million years of evolution. The theory behind it is more fully explained in *The Penguin Encyclopaedia of Nutrition*. The diet allows you to eat as much as you like of meat, fish, eggs, leafy vegetables, butter, margarine, cream, or any oil or fat. It recommends that you take up to half a pound of fruit a day, and one pint of milk. You are given a list of the carbohydrate content of foods and drinks in units of five grams, which I call Carbohydrate Units, and you are told to take about ten of these in a day.

My interviews with overweight patients begin with general questions about health, and some of these are about indigestion: "Do you have indigestion, or any sort of pain or discomfort after meals? Where do you have the pain? What sort of pain is it? How often do you have it? How long does it last? What do you take to relieve it?" After lots of other questions about their health, the patients are examined and weighed and measured. After a few weeks of repeated visits by the patient I go back to these questions and I find, for example, that, having lost some weight, they are not so short of breath, not so tired, have no pains in their hip joints, no longer suffer from swollen ankles at the end of the day.

All of these changes I expect, but as I first noticed years ago, many of them also said, with surprise, that my questions reminded them that they had stopped having indigestion. And this relief was observed not just after they had lost weight but almost from the moment they had begun the low-carbohydrate diet.

Let me interpose my personal experience. When I was young, I suffered from severe dyspepsia, and was diagnosed as having a duodenal ulcer. I was given what was then very up-to-date advice: not to have an operation unless it became imperative, to continue with my work, to "take it easy" and not get too exhausted, and to avoid spicy foods, eat more frequently, and eat small meals. I gradually gave up cakes and pastries too, because I found I always got heartburn after these foods. But I still had quite frequently to take antacid preparations such as magnesia or aluminum trisilicate.

I later discovered that, like many very sedentary middle-aged men, I was beginning to put on weight. Obviously, I now reduced my carbohydrate intake very considerably, as I had advised my patients to do, and this got my weight under control. Suddenly, a few months later, I became aware that my indigestion had almost entirely disappeared.

On the strength of these observations I decided to set up a proper test of the idea that a low-carbohydrate diet really does relieve the symptoms of indigestion. This was a more formidable undertaking than you might think.

Severe indigestion often occurs in people who are under a great deal of stress, and so are not always very reliable in their statements. Secondly, indigestion often comes in bouts—a few weeks of pain and then, for no apparent reason, a few weeks or even months with no pain at all. If you happen to be taking some treatment—any sort of treatment—before you have one of these intermissions, then you are likely to believe that it was the treatment that made you better. Thirdly, no doctor has a certain and objective measure of how much pain other people are experiencing; you have to accept their own estimate of whether this indigestion is better or worse, and if so whether it is slightly or considerably better or worse.

Nevertheless, I thought it was worth attempting to see whether a low-carbohydrate diet did improve the symptoms of dyspepsia. So we set up a fairly comprehensive scheme of experiments that would eliminate the difficulties, or at least minimize them. The tests were carried out at King's College Hospital in London. Physicians and surgeons were asked to send us anyone coming to them complaining of severe dyspepsia that had lasted, though perhaps not continuously, for more than six months. Many had had symptoms for five years or more. The only patients not included in our experiment were those who were going to have an operation for their condition.

Each patient was carefully questioned and examined by a physician, and than sent on to a nutritionist. Alternate patients were instructed either in the conventional dietary treatment commonly used then, or in the low-carbohydrate diet. The conventional treatment consists of telling the patients to avoid fried foods and irritants such as pickles or foods containing spices, to take frequent small meals, and to avoid alcohol, especially on an empty stomach. At intervals each patient came back to the physician for assessment of the progress of his condition, and to the nutritionist to check the diet that he was following. The physician did not know which diet each patient was on; the nutritionist did not know how the patients were progressing.

After three months, the diets of the patients were reversed, so that those taking the conventional diet were transferred to the low-carbohydrate diet, and those taking the low-carbohydrate diet transferred to the conventional diet. The experiment then continued for a further three months.

Having made the conditions of the experiment so stringent, we were not surprised that it took us more than two years to get together information on 41 patients who had reported regularly for six months, and had, as best we could judge, adhered to our instructions. From the detailed records kept by the physician, he and I then separately assessed their total progress and classified them as having shown no change, or having reported various degrees of improvement or deterioration at the end of each three-month period. Our assessment differed in only one or two instances as to the degree of change, but not once did we disagree as to whether the patient reported that he was better or worse or just the same. It was only after the clinical assessment that we looked to see whether the patient had begun with the low-carbohydrate diet or with the alternative diet.

In summary, the results are pretty clear. Of the 41 patients in our trial, two said that they were worse on the low-carbohydrate diet, 11 said that they were no different on either diet, but a decided majority—28—said that they were very much better on the low-carbohydrate diet. Some of these were quite certain that the improvement was so great that nothing on earth was going to make them give up the low-carbohydrate diet. One said, "I feel better than I have been for five years." Another was even more enthusiastic: "I have never felt better round my stomach in all my life." The patients included men and women, some with gastric or duodenal ulcers, some with hiatus hernia, and some who probably had ulcers which, as so often happens, had not been revealed by X-ray examination.

These results, of course, pleased us a great deal. They suggested that chronic and severe indigestion, from several causes, could be greatly relieved by diet alone in something like 70 percent of patients. This result was especially pleasing because there had been increasing disappointment in the last few years about the results of dietary treatment of these conditions. Several research workers had put patients on fairly strict "gastric diets"—steamed fish, white meat, mashed potatoes, milk puddings—or on the more liberal but still fairly conventional diet I described earlier. All these investigators had concluded that the diets did not seem to relieve the severe dyspepsia of their patients, whether or not they had a definite ulcer. Now it can no longer be said that diet does not relieve severe dyspepsia. The right diet may well do so; but of course the wrong diet will not.

The low-carbohydrate diet we had used in our study was limited in both starch and sugar. For a variety of reasons we suspected that it was the reduction in sugar that was chiefly responsible for the improvement we saw, so we carried out a further experiment to look at the effect of sugar in a normal diet. Working with young men, we managed to persuade seven of them to swallow a gastric tube first thing in the morning. They did this before and again after two weeks of a high-sugar diet. Through this tube we obtained samples of their gastric juices at rest, and then further samples were taken at 15-minute intervals after they had swallowed a bland "test meal" consisting mainly of pectin. Each sample was analyzed in the standard ways, most importantly by measuring the degree of acidity and digestive activity.

The results showed that two weeks of a sugar-rich diet causes an increase in both acidity and digestive activity of the gastric juice, the sort of change you often find in people with such conditions as gastric or duodenal ulcer. The sugar-rich diet increased the acidity by 20 percent or so; the enzyme activity was increased nearly threefold. And let us remember that these effects were seen early in the morning, before breakfast—two weeks on the high-sugar diet had made the gastric mucous membrane much more sensitive to the very mild stimulus of the pectin test meal.

Peptic ulcer

Our experiments on indigestion were carried out in the days before the new drugs cimetidine and ranitidine became available for the treatment of gastric and duodenal ulcers. Nine of the dyspeptic patients we had treated, of whom six had improved on our diet, had been diagnosed as having one or other of these peptic ulcers. Today, these drugs are used to give good and prompt relief to most such patients, and their ulcers usually heal. Although the symptoms are likely to recur, they will probably be relieved again by the resumption of drug treatment. Strict diets are therefore now used far less frequently than they were for the treatment of either sort of ulcer.

Nevertheless, there are always disadvantages in long-term drug treatment. Although cimetidine and, especially, ranitidine are unlikely to produce side effects, they do sometimes do so. Moreover, there are now more and more doctors, and especially patients, who are reluctant to embark on a course of drug treatment that may continue indefinitely, even though with intermissions. It is my own opinion that patients should be encouraged to try a low-carbohydrate diet before the decision is taken to use drug therapy.

Patients with duodenal ulcer have been shown to have a diminished glucose tolerance and an increased blood insulin: two of the features produced by a diet high in sugar.

Hiatus hernia

The type of dyspepsia that responds perhaps most strikingly to the low-carbohydrate diet is hiatus hernia. To understand this condition, you have to picture the esophagus (gullet) passing through the diaphragm as it leaves the chest and enters the abdomen to join the stomach. If there is a weakness in the diaphragm near where the esophagus passes, then pressure in the abdomen, from whatever cause, may push part of the abdominal esophagus and the adjoining portion of the stomach back through the weak part of the diaphragm. The usual symptoms are heartburn occurring soon after a meal, a feeling of excessive fullness of the stomach, and often a severe pain. Since food cannot easily pass into and through the stomach, some of its contents may pass back into the esophagus. The acid from the stomach can now irritate the esophagus and result in what is called "reflux esophagitis." The pain occurs mostly at night, and is considerably relieved if the patient sits up.

The usual treatment includes advising the patient to eat small, nonirritant meals consisting of bland foods, like the so-called "gastric" diet. The last meal should be taken early rather than just before going to bed. The patient should avoid bending, lifting, or straining, and should reduce excessive body weight; sleeping with a raised pillow can also help to ward off the pain.

Most of this advice is worth following. However, our experience indicates that an even better diet is one in which the carbohydrate is considerably restricted and sugar virtually eliminated. Many patients have reported that, having adopted this diet, they found significant relief for the first time.

Gallstones

One of the conditions that often shows itself simply as indigestion is gallstones. The stones, which almost always contain a high concentration of cholesterol, frequently collect in the gall bladder, where they produce inflammation, or cholecystitis. Gallstones are said to be present in 20 percent or so of adults (and rather more frequently in women than in men), but about half of these people never have symptoms. However, according to one up-to-date book of medicine, "In prosperous countries the incidence of symptomatic gallstones seems to be increasing, and occurring at an earlier age."

Those individuals who do have symptoms of gallstones, and have consequently been investigated, have often been found to have one or two additional features. These include Type II diabetes, hiatus hernia, an increased blood concentration of triglyceride and insulin, and obesity; for example, patients with gallstones weigh on average 5.5 kg more than patients with no symptoms of the disease. All this makes us think that, once more, we are dealing with a disease in which dietary sugar may be involved. This suggestion was reinforced by the fact that one of the patients whose dyspepsia was considerably improved in our trial of the low-carbohydrate diet was a patient whose dyspepsia had been diagnosed by her doctor as being caused by gallstones. This left us with the thought that perhaps her gallstone disease as well as her indigestion had been caused by her usual diet with sugar.

Since we completed that study, some New Zealand research workers have reported finding that patients with gallstones tended to be taking more sugar than did people of the same age, sex, and occupation who did not have gallstones; they also had a higher concentration of insulin in the blood. Their report refers to 124 men and 219 women with gallstones, whom they compared with 111 normal men and 211 normal women. The results showed that the people with gallstones, both men and women, took more sugar, chiefly in beverages and confectionery, than did the control subjects. The authors calculated that an increase in daily sugar consumption by 40 grams—equivalent to two spoons of sugar in each of three or four cups of tea or coffee—more than doubled the risk of the individual developing gallstone disease.

Again, other research workers found that sugar in the diet could produce gallstones in hamsters and in dogs. The latest research, in England, has shown that vegetarians are less likely than meat-eaters to have gallstones. This could be due to something in meat that promotes gallstones, or something in vegetables that prevents their formation. But it may well be due to the fact that vegetarians tend to take less refined sugar and less sugary foods and drinks.

Crohn's disease

Crohn's disease is an unpleasant condition of the alimentary canal which chiefly affects men and women between the ages of 20 and 40. Its chief characteristics are bouts of pain with diarrhea; the pain may be so severe as to mimic appendicitis. It can affect any part of the digestive tract. No one knows the cause of Crohn's disease, and there is as yet no satisfactory treatment. Occasionally, it is necessary to remove the part of the bowel that is particularly badly affected.

In a study in Bristol, England, 30 patients with recently diagnosed Crohn's disease were asked about their usual diet before they developed the disease. These diets were then compared with the diets of 30 healthy people matched for age, sex, and social class. The patients were found to have been taking 122 grams of sugar a day on average, compared with 65 grams for the control subjects. Their dietary fiber intake was slightly lower at 17.3 grams compared with 19.2 grams. In other respects the diets were much the same for patients and control subjects.

The Bristol doctors then advised their patients to take a diet that was rich in fiber and low in sugar. They compared these diets and the patients' responses to this treatment over an average of 52 months with the diets and responses of a carefully matched series of patients who had attended the same clinic in previous years.

The results showed that the current patients had been admitted to hospital for an average total of 111 days during the 52-month study, compared with an average of 533 days for the patients who had not been on the new diet. The sugar intake of the current patients had been reduced to 30 grams a day, as compared with 90 grams for the non-diet-treated patients.

These findings have been confirmed by a similar study in Italy of 109 patients with Crohn's disease. There it was calculated that a diet with a high sugar content increased by two and a half times the risk of developing the disease.

The Italian doctors also examined the diets of people who developed ulcerative colitis. This is a condition with some resemblance to Crohn's disease, except that it affects only the large intestine (the colon), and there are no strictures in the bowel, but ulcers. These may become so severe and so deep as to perforate the bowel. The main symptom is bouts of severe diarrhea with blood and sometimes pus in the stools. In some patients it is difficult to distinguish whether they have ulcerative colitis or Crohn's disease. The Italian study was concerned with 124 patients. From an examination of their diets, the investigators calculated that here too a high consumption of sugar increased the chance of developing ulcerative colitis to two and a half times that of persons with a low sugar consumption.

17

A host of diseases

I now want to talk of a number of quite unrelated conditions in which there is evidence of very varying strength that sugar might perhaps be involved.

Damage to the eyes

Ophthalmologists had for a very long time wondered whether nutrition could affect the way the eye developed, and thereby affect such conditions as long-sightedness or short-sightedness. There was some suggestion that short-sightedness (myopia) occurred in children when their diets were short of protein. The research on which this notion was based was not considered acceptable by the experts, and there is nowadays no support for this view. One of my colleagues, together with an ophthalmologist, looked at the problem by doing some experiments with rats. They too could find no effects of diets that were simply deficient in protein.

They then studied the effects of diets that were low in protein but also high in sugar, the types of diets known to be common among the rapidly increasing populations of the large cities in the poorer parts of the world. In one experiment, they fed rats on diets low in protein and with or without sugar. After six or seven months, both of these groups had grown poorly compared with control rats fed the normal high-protein diet. The investigators found no significant difference in refraction between the normal rats and those on the low-protein high-starch diet, but the rats fed the low-protein high-sugar diet had a considerable degree of myopia, amounting to one diopter.

In a second experiment they took another three groups of rats: one group was placed on a normal diet; the second on the low-protein diet with sugar; the third group on the normal diet but with the amount restricted so that the rats grew at the same low rate as did the rats on the sugar diet. After nine weeks there was no difference in refraction between the groups. But by 15 weeks the sugar-fed rats had developed myopia, again to the extent of nearly one diopter, compared with the normal group and with the poorly fed group.

At this point the diets of the second and third groups were reversed. One result was that the poorly fed group, with normal refraction up to the time of the change, became myopic within three weeks of having started the diet with sugar. The other result was that the sugar-fed group with myopia at the time of the change-over did not improve during the whole of the rest of the experiment, even though it lasted for 23 weeks after the change.

We also measured eye refraction in student volunteers who were the subjects of one of our experiments. As before, we took a large number of measurements before and after they were given a high-sugar diet. After two weeks on this diet there was a small but quite significant change in their refraction—but this time it was a change toward long-sightedness, not toward myopia or short-sightedness.

At present, we are suggesting that the reasons have to do with the level of glucose in the blood. Doctors have known for some time that diabetics develop a mild but noticeable degree of short-sightedness if their blood sugar is not properly controlled and consequently rises to an unduly high level. We believe that this may be the cause of the myopia occurring after a long period in our rats on the high-sugar diet; we know that such animals become mildly diabetic with a high blood sugar and that a low-protein diet probably accentuates the condition. In our students, the two weeks on a high-sugar diet tended to produce a low blood sugar, as I have shown, so one would have expected not myopia but long-sightedness.

I have already mentioned (p. 123) that severe changes occur in the retina of the eye in diabetes. And I pointed out that similar changes can be produced in rats by feeling them with sugar.

Damage to the teeth

Each year, millions of teeth are extracted by dentists from children all over the Western world. In the UK alone, the loss is four million teeth weighing more than four tons. In one survey in Dundee in Scotland, 13-year-old boys and girls were found to have an average of ten decayed teeth. More than one third of British adults over 16 have had every one of their teeth extracted.

Fossil evidence suggests that the condition now known as "dental caries," or "dental decay," hardly occurred in prehistoric times, before the introduction of agriculture and the great increase of starchy foods like cereals in man's diet. The disease became much commoner only recently. There is no doubt that this is associated with the introduction of sugar as an increasing component of the conventional diet.

To understand the process of dental decay, we should know a little about the structure of the tooth. It is mainly made up of dentine, a sort of tough bone. This is covered by a thin layer consisting of enamel, the hardest tissue in the body. Inside the dentine is the soft pulp from which the dentine is made, and in it are blood vessels. As anyone who has had a toothache or visited the dentist knows, the pulp also contains highly sensitive nerve endings.

Dental decay begins from plaques of material that stick on the surface of the teeth, and are found especially in the normal fissures and crevices of the tooth surface. The plaque is made up of a background material of protein and carbohydrate, which retains particles of food, debris from the saliva, and countless bacteria.

Present evidence is that dental decay proceeds by the production of acid by bacteria in the plaque, especially bacteria belonging to the type called *Streptococcus mutans*. The acid is produced in the plaque as it adheres to the surface of the tooth. It is not washed away by the saliva but gradually attacks the dentine until, unchecked, it exposes the sensitive pulp. The production of acid is facilitated by the build-up of a complex carbohydrate in the plaque.

It seems that what the acid-producing bacteria like best in the plaque is the particular complex carbohydrate called dextran. This can be built up from any sugar chiefly by the streptococci, but very much more is produced from sucrose.

Some people are less susceptible to caries than others, partly because they appear to have inherited a higher resistance than normal, partly because they live where the drinking water contains adequate amounts of protective fluoride, partly because they clean their teeth frequently, but chiefly because they do not consume food and drinks that allow their teeth to come into prolonged contact with sugar. The epidemiological evidence about the development of dental decay includes that which I summarized earlier in regard to primitive human beings. I showed that carbohydrates in general are a relatively recent addition to the human diet. Perhaps the earliest specific mention of the association between sugar and caries was that of a German traveler who in 1598 remarked on the black teeth of Queen Elizabeth of England, "a defect the English seem subject to from their too great use of sugar." Much earlier, Aristotle spoke of teeth being damaged by figs, but he was of course not aware that their sweetness was largely due to the same sucrose that was later extracted from the cane and manufactured into the sweetmeats that ruined Queen Elizabeth's teeth.

Dental caries has become the scourge of the wealthier countries mostly during the present century. Here are some examples. In the UK, 26 million teeth were filled in 1965; in 1983 the figure was 40 million. In America, as recently as 1980, it was found that 17-year-old teenagers had an average of six carious teeth. In Norway, in 1981, there were 16 carious tooth surfaces in the average 16-year-old. In New Zealand schoolchildren had an average of 1.5 teeth filled during 1980. And it is said that in Germany only 0.1 percent of children over 15 have caries-free teeth—that is, only one in a thousand.

Percentage of children in England and Wales with tooth decay

Age	1973	1983
5	71	48
8	91	73
10	93	80
15	97	93

There has been a fall in the amount of tooth decay in British children in recent years, as you can see in the table. This is likely to have been due to the increase in the number of areas where water is fluoridized, the increase in the availability of fluoride toothpaste, and perhaps a slight response to increasing publicity about how to preserve the teeth. Nevertheless, there is no room for complacency when the average 15-year-old in the UK in 1983 has more than five decayed, missing, or filled teeth. In particular, the decrease in tooth decay has occurred only in the industrialized countries; in the developing countries, the rapid rise in caries was described by the WHO as "absolutely frightening."

Number of decayed, missing, or filled teeth of children in England and Wales

Age	1973	1983
7	0.8	0.8
10	3.0	1.6
15	8.4	5.6

The increase of tooth decay in populations such as the Eskimos and the inhabitants of Tristan da Cunha followed their introduction to sugar. The increase in the developing countries parallels in general the rise in the availability of sugar to these populations. In the other direction, children in Western Europe experienced a reduction in dental caries during and soon after the two world wars when sugar was scarce, but its prevalence increased rapidly when sugar became freely available again.

If we are to get rid of dental caries altogether, it is unlikely to occur simply from a continuation of the sizable reduction that, it appears, has come largely from fluoride. It will be necessary also to persuade people, and children in particular, that they should avoid eating sugary foods, especially sweets and chocolate that tend to stick to the teeth. And ideally this would be part of a program of nutrition education. But the problem of general nutrition education is not as easy to solve as it seems. Long ago my colleagues and I became increasingly aware that it is necessary not only to teach people facts about nutrition, but also to get them to use these facts. More simply, the purpose of nutrition education of the public is not just to improve nutritional knowledge, but to improve nutritional behavior.

A small piece of research that we carried out illustrates this very well. We asked a hundred or so London mothers some questions about healthy eating, one of which was, "What is the chief reason why children get holes in their teeth?" More than 90 percent of the mothers answered that it was because the children ate sweets. But knowing this did not prevent them from buying sweets for their children. When, therefore, I was asked to give the Annual Foundation Lecture at Newcastle Dental School, I chose as my title, "Dental caries is preventable: why not prevented?," and discussed the whole problem of nutrition education for the public. I especially made the point that almost everyone knew that a major cause of tooth decay was the eating of sticky, sugary confectionery, cakes, and biscuits, yet that did not stop children buying these items or being given them by their elders.

The lecture was published in the *British Dental Journal* and evoked a very angry letter from Professor B. Cohen, who was doing dental research at the Royal College of Surgeons in London. He thought it ridiculous that I had not pointed out that the holes in the teeth were caused by bacteria that produced acid. He wrote: "Until it is accepted that caries is a disease caused not by sugar but by the action of bacteria on sugar, effort will continue to be expended in preaching deprivation that few patients will ever practise, instead of striving to devise means for the management of an infection." Professor Cohen was at that time carrying out research designed to see if it was possible to produce a vaccine against the "infection" of the mouth with the caries-causing *Streptococcus mutans*. The experiments were with monkeys, which were encouraged to develop tooth decay; the efficacy of the experimental vaccines was tested by injecting them into half of the monkeys. I visited the laboratory soon after my lecture, and I don't suppose I need to tell you how dental decay was induced in the monkeys. You'll have guessed, I am sure, that it was by giving them lots of sticky sweets.

My talk in Newcastle took place in 1969, some eighteen months after the birth of my grandson, and a few months before the birth of my granddaughter. They haven't had anti–*Streptococcus mutans* vaccine; they have, though, been careful about eating sweets and chocolates. At the ages of 18 and 16, he has one filling and she has none. As I mentioned in Chapter 3, my grandson refused to eat his birthday cake when he was three, because it was too sweet.

These observations do not prove that sugar is a cause of dental decay. I have already pointed out that the association of disease and diet in different populations can only be taken as a clue to the cause. Next, one must see whether the individuals in any one population who get caries are those who eat a lot of sugar. Curiously enough, not much research of this sort has yet been done. The dentists in Dundee to whom I referred earlier examined 13-year-old boys and girls in 1960, 1961, and 1962. They found more dental caries in those that ate more sweets, but surprisingly they found no difference in caries among those that did or did not brush their teeth regularly.

Our own research, carried out in 1967 on a much smaller number of children, also showed that there was more decay in children taking more sugar in solid foods (that is, more sweets, biscuits, and so on). But we also found that the strong relationship between sugar and decay occurred only in children who did not clean their teeth regularly; if they did clean their teeth regularly, they had little caries even when they ate a lot of sugary foods.

Many experiments have been done, especially in animals, to see what changes in diet affect the teeth. As always, the precise results differ according to what animals were used, exactly what the experimental diets were, how the diets were given, and for how long. The general results, however, seem clear. When there is no carbohydrate, little or no caries is produced. Diets containing starch, or bread (brown or white), produce either the same amount of caries or a very little more. Diets with any sort of sugar produce much more caries, and the most "cariogenic" sugar is sucrose.

The best-known experiments with children are those done by the British Medical Research Council in 1950, and in the town of Vipeholm in Sweden a few years later. The first study lasted for two years and showed that the addition of sugar during mealtimes did not increase the amount of caries in children. The second study compared sugar given in different ways for four years, and found that little additional caries occurred if the sugar was taken at meals, but much more occurred when it was taken as sweets between meals, and especially if it was taken as sticky toffees between meals. Obviously, what matters is whether the sugar is in contact with the teeth for some time. Sticky sweets and cakes and biscuits between meals are the chief culprits, especially if their residue is allowed to remain without being exposed to a good and prolonged tooth brushing.

A great deal of attention has been directed toward what is called "rampant caries." The custom seems to have grown of giving babies dummies to suck that have a small container in which syrup is put. The effect of this, or of giving ordinary dummies constantly dipped into sugar, is that the babies' teeth become rotten as they erupt, so that at the age of two or three years their mouths are full of blackened stumps. In one survey, one baby in 12 was found to be suffering from rampant caries; in another, the figure was one in eight.

One of the most interesting and unexpected of the observations of the role of sugar in producing dental caries comes from a study of a rare disease, hereditary fructose intolerance. Only a few families have been discovered with members suffering from this disease, and they become violently sick whenever they get fructose, or sucrose, which you will recall is a compound made up of equal amounts of glucose and fructose.

Very early in life, therefore, they learn to avoid fruits and anything containing sucrose. They can and do eat starchy foods, since starch is digested to give only glucose. But even though they eat lots of white bread, made from what people like to call "refined flour," they have very little caries, and what they do have is of a very minor degree.

One day perhaps we shall be able to immunize children against the bacteria that are involved in producing caries. But although experiments aimed at producing immunity to caries have been in progress for nearly 20 years, a practical vaccine has still not been produced.

Damage to the skin

In measuring the amount of sugar consumed by hospital patients, I was chiefly interested in those with coronary disease. But it occurred to me that it would be interesting to see how much sugar was taken by patients with two or three other conditions. There are, for example, conditions such as acne (blackheads) which occur quite frequently in teenagers, and which doctors believe is caused or made worse by the eating of confectionery.

We measured sugar intake in these patients and compared it with that of people of the same age and sex without acne. We also decided to look at another common skin disease called seborrheic dermatitis, but not this time because diet had been implicated by physicians. The reason was that this condition has to do with the secretion of the glands in the skin of the oily substance called "sebum." There is some evidence that this material is altered when the diet is rich in sugar. So we also measured the sugar intake of patients with this disease, and compared it with that of people without seborrheic dermatitis, each one chosen so as to be the same sex and age as a patient with the disease.

It turned out that the acne patients were not taking any more sugar than the control subjects, but that those with seborrheic dermatitis were taking appreciably more.

The implication of these results is that sugar is not involved in producing acne, but may be involved in producing seborrheic dermatitis. We could extend these conclusions by saying that it is unlikely that acne patients would get better if they ate less sugar, although it may be that they are especially sensitive. They might therefore be suffering from acne partly because of sugar, even though they do not take more than other people, and if this were so they would indeed be better off taking less sugar. But no one has really done a properly controlled test to see whether less sugar does make them better.

As for seborrheic dermatitis, the fact that sufferers are heavy sugar-eaters at once suggests that we should see if we can improve them with a low-sugar diet. Although the results look promising, we have not been able to continue our research; we shall have to wait for others to take it up.

Damage to the joints

Gout has always interested doctors. The popular idea of gout is that it is found in people who overindulge in rich food and in alcohol; in England, we think of the retired colonel drinking his bottle of port a day. It is often thought to be very rare nowadays, but in fact it is not all that rare. It occurs mostly in middle age and later, and more in men than in women.

The reasons that made it seem worth looking at sugar consumption by gouty patients were pretty flimsy, I must admit. First, one of the features often found in people with atherosclerosis, and found in all people with gout, is a raised level of uric acid in the blood. Second, in human beings and in some animals a diet high in sugar increases the concentration of uric acid in the blood. Third, people with gout are more likely than other people to get coronary thrombosis, and, conversely, people with coronary disease are more likely to have gout.

So we studied patients in two or three rheumatic clinics. We had in fact three groups of people: patients with gout, patients with a different rheumatic disease, rheumatoid arthritis, and normal individuals who were apparently quite healthy. As usual, the three groups of subjects were matched for age and sex.

As we half expected, the patients with rheumatoid arthritis were eating the same amounts of sugar as the control subjects. But the patients with gout were taking appreciably more sugar than the control subjects; the median values were 103 grams of sugar a day for the gouty patients and 54 grams for the control subjects.

Disease of the liver

The liver is the most active organ in the body; every item of food and drink that has been digested in the alimentary canal and absorbed into the blood goes straight to the liver. There the large vein carrying the blood from the alimentary canal breaks up into capillaries that bring the products of digestion close to the liver cells. In these cells, most of the very varied substances resulting from digestion and absorption undergo chemical transformation into materials that are to be used by all the different organs in the body—including the liver itself—to repair their wear and tear, or to be metabolized as fuel, or to be stored for future use. In addition, the liver has the task of detoxifying several of the harmful materials that may have been present in the food or produced during metabolism. For all these reasons the liver is one of the first organs to be affected by undesirable items in the diet.

Many of the activities of the liver are affected not only directly by the arrival of varying quantities of these substances in the blood flowing through the liver, but also indirectly by changes in the hormone content of the blood. And since, as we saw, the amounts of at least three hormones are affected by sugar in the diet, we were keenly interested to look at some of the changes that sugar produces in the liver.

I have already mentioned (p. 107) the enlargement of the liver that is produced by sugar. Part of this increase in size is caused by the accumulation of fat from the liver cells; in some instances the amount of fat is enough to make the liver yellowish in color, just like the diseased "fatty liver" that occurs occasionally in people who are poisoned with such materials as chloroform, or in alcoholics.

When my colleagues investigated whether part of the enlargement of the liver was due to an enlargement of its cells or to an increase in the number of cells, it turned out that the effect of sugar is to increase both cell size and cell numbers. Although to some extent a technical matter, an increase in the number of liver cells shows that some of them have actually divided, indicating a more profound action of sugar than simply to make cells swell up.

Most recently we have been working closely with our colleagues in the Biochemistry Department at Queen Elizabeth College to study in more detail the changes that sugar produces in the liver. One of the reasons for this particular study was a report, as long ago as 1949, that not only alcohol but also sugar can produce fibrosis of the liver—that is, an increase in the sort of "scar tissue" that precedes the development of cirrhosis of the liver. This research was carried out by a group of scientists led by Dr. Charles Best, who was one of the people responsible for the discovery of insulin in 1921. Other researchers have repeated the work of Best and his colleagues and produced the same results.

All of this earlier work was done with rather special diets lacking particular nutrients. That is why the research team at Queen Elizabeth College has been looking at the effects of our routine diets, which are not deficient in any obvious way, and differ only in whether or not the carbohydrate part of the diet includes sugar. The most recent experiments have used extremely sensitive biochemical analyzes to detect in the blood and in the liver the chemical fragments that the body uses to build up the collagen that is increased in liver fibrosis. Collagen, which exists in several forms of slightly different chemical structure, is the protein that is present in the walls of the body's cells, and also makes up a great part of the connective tissue that exists in the sinews or tendons, in cartilage and bone, and in scar tissue. Our research has revealed that in sugar-fed rats there is a distinct increase in these fragments, both in the blood and in the liver, long before it is possible to detect fibrosis in the liver with the microscope. The same increase is seen in rats with diabetes and in human subjects with cirrhosis of the liver caused by chronic alcoholism.

Is there a link between sugar and cancer?

There are some cancers that appear to have become more common in the last 50 or 60 years, and that also appear to be more common in the affluent than in the poorer countries. So I thought it might be worthwhile to see whether there was any relationship between the numbers dying from these cancers in several countries, and the amounts of sugar that their populations consumed.

The usual snags faced us with the epidemiological studies. How many countries are there that keep proper records of the causes of death in their populations? Even where records are kept, how sure can you be that the diagnosis of cancer is correctly made, or made on exactly the same criteria, in different countries?

Some sorts of cancer can be diagnosed fairly readily; others are often misdiagnosed. Because of this, we gave most attention to three or four where the experts tell me that there is a reasonably good chance of correct diagnosis.

The evidence at present comes chiefly from a study of international statistics and takes the form of an association between the average sugar consumption in different countries, and the incidence of two or three particular forms of cancer. The cancers that seem most likely to be related to sugar consumption are cancer of the large intestine in men and in women, and cancer of the breast in women. The death rate for these three cancers in different countries is quite closely associated with average sugar consumption, to about the same extent in fact as the association between sugar consumption or fat consumption with the death rate due to coronary disease. An example is found in the international statistics for 1977–79 for breast cancer deaths in women over 65. The five countries with the highest rates are, in descending order, the UK, the Netherlands, Ireland, Denmark, and Canada; the highest levels of sugar consumption, again in descending order, are in the UK, the Netherlands, Ireland, Canada, and Denmark. On the other hand, the lowest mortality, in ascending order, is in Japan, Yugoslavia, Portugal, Spain, and Italy, with the lowest sugar consumption in Japan, Portugal, Spain, Yugoslavia, and Italy.

My own observations of the association between sugar and cancer of the intestine and of the breast were made several years earlier than the study I have just quoted. I calculated what are called the "correlation coefficients" between these cancers and sugar consumption in all the countries for which statistics were then available.

Let me explain first what correlation coefficients are, and let me take as an example the relation between people's height and weight. On the whole, the taller people are, the more they weigh. But it is all very well to say that there is "on the whole" this association between height and weight; it would be better if we could say how close this association is. Supposing that it was a precise and exact association, so that the person who was only a little taller than another would inevitably be heavier, and one still taller would be still heavier. If this were so, you would say that the correlation coefficient was 1.0.

Supposing on the other hand—and this is even more unlikely—that there was no relationship whatever between height and weight, so that it would be just as likely for a man weighing 150 pounds to be five feet tall or six feet tall. In this case the correlation coefficient would be 0. In fact, there is a relationship, but not a precise one; tall people *tend* to be heavier. If you work it out exactly, for adult men the correlation coefficient between height and weight comes to about 0.6.

The correlation coefficients I have found so far for cancer and sugar consumption in different countries are as follows:

Cancer of the large intestine in men: 0.60
Cancer of the large intestine in women: 0.50
Cancer of the breast: 0.63

However, such international statistics, as I have stressed repeatedly, can do no more than give a clue as to the possible role of sugar or fat in producing disease. But there are indirect studies that suggest why sugar might increase the risk of women developing cancer of the breast, and of both men and women developing cancer of the large bowel.

During the past ten years the development of cancer of the breast has been linked with the female sex hormones, especially estrogen. Evidence for this association comes from studies in several different countries. It has been suggested that cancer of the bowel may be caused by a high concentration of insulin in the blood, but a more likely possibility, put forward by workers in America, Hawaii, and England, is that, like breast cancer, it is estrogen that is involved. Again, there has recently been an increase in the incidence of testicular cancer in young men, and it has been shown that their mothers were often overweight during pregnancy, and have an increased concentration of estrogen in their blood.

Whatever is ultimately shown to be the cause of these cancers, the fact is that a high consumption of sugar can produce an increased blood concentration of both of these hormones—insulin and estrogen.

There have also been suggestions that other dietary components might be involved in causing cancer, especially of the bowel. One suggestion blames a lack of dietary fiber; a second is that it is caused by an excess of saturated fat and a deficiency of polyunsaturated fat. I have already explained why, on evolutionary grounds, dietary fiber, especially from cereals, and polyunsaturated fat, largely from oil seeds, are not likely to have been significant contributors to the diets of our ancestors, long before there began to be a high prevalence of the diseases of affluence. Moreover Japan, with a very low incidence of bowel cancer, has a fiber intake very similar to that of the UK. And, experimentally, it is polyunsaturated fats rather than saturated fats that tend to produce cancer in animals.

Sugar and drug action

Let me here add one more interesting effect of sugar, although no one has yet pursued it sufficiently to see if it has any practical application. A dozen or so years ago, some of my colleagues looked at the question of whether diet has any effect on the action of drugs. They gave a common sedative drug, pentobarbitone (Nembutal) to rats that had been fed on a diet in which the carbohydrate was either starch or sugar, and they recorded how long the animals slept after taking the drug. They found that the rats taking the sugar diet slept for a significantly shorter time than did the rats taking the starch diet—on average 98 minutes compared with 141 minutes.

This discovery raises several questions. Would sugar decrease the effect of Nembutal in human beings as well as rats? If sugar has this effect, do other dietary components behave in the same way? Can you deliberately decrease—or increase—the effects of other drugs by changing the diet? If so, would the effect be produced simply by changing the amount of sugar or other dietary component for just a day or two before the drug treatment begins, or even at the time it begins?

Clearly, there is a lot more research to be done by nutritionists and pharmacologists before we have the answers to these questions.

Sugar and protein

One of the unexpected effects that we found when putting sugar into the diet was that it interferes with the body's use of the protein in the diet. We first noticed this when we fed our rats a diet that was low, but usually adequate, in protein. When fed on our normal diet with starch the rats grew perfectly well, but when we replaced the starch with sugar their growth was retarded. Looking more closely at the reason for this, we found that the sugar interfered with the body's use of protein, so that the rats lost protein instead of accumulating it as they should when growing.

Protein utilization is measured in an indirect way. You measure the nitrogen in the diet, since virtually all of this is in the dietary protein; you also measure the amount of nitrogen in the urine, which is where the body gets rid of the products of protein metabolism. If the dietary nitrogen is more than the urinary nitrogen, the body is retaining some protein, and this we call a positive nitrogen balance. If the dietary nitrogen is less than the urinary nitrogen, the body is losing protein—that is, the body is in a negative nitrogen balance.

Here then are the results of three of our experiments with rats:

Effect of sugar on protein utilization

Diet	Exp. 1	Exp. 2	Exp. 3
With no sugar	+17	+17	+6
With sugar	-2	+2	-6

This work with animals suggests that a diet low in protein may be even more deficient when it is accompanied by sugar. It seems that the usefulness of a given amount of protein—its value in promoting growth—is diminished when sugar is present in the diet. While there is no direct evidence that what is true of rats and chickens is equally true of growing children, this interrelationship could be of particular interest in poorer countries.

One of the characteristics of such countries is the enormous increase in urbanization. There is a tremendous influx of people from the country into the big cities of India, Thailand, South America, Ghana, Nigeria, and so on. The chief effect on the diets of the new arrivals, mostly extremely poor, is an increase in the consumption of manufactured foods such as cakes and biscuits and soft drinks, so that they take even less protein than before, but more sugar. If the effect in children is similar to that in young animals, the combination of sugar and low protein would explain even better the high incidence of protein deficiency than would the low protein alone. It is protein deficiency that causes the dreaded disease kwashiorkor that is so common, and often fatal, in the developing countries.

A high intake of sugar is more usually a feature of diets in well-off countries, and these are more likely to be adequately supplied with protein and usually with other nutrients too. The question has been raised whether this high intake of sugar causes an increase in growth, rather than a decrease, and we shall look at this in the following chapter.

A wide range of disorders

This is a very mixed collection of diseases, to be sure—cancer, dental caries, short and long sight, dermatitis, and gout, as well as coronary disease, diabetes, and various digestive conditions. And the evidence that they are caused in part by an excessive consumption of sugar is by no means equally convincing for all of them. At one extreme, it seems that everybody is certain about the role of sugar in dental caries, except perhaps the manufacturers of biscuits and confectionery. At the other extreme, there is as yet not much evidence that cancer of the colon or of the breast is really any more likely to appear in people who eat a lot of sugar. I shall, however, be content if you will agree that for cancer especially, but also for gout and seborrheic dermatitis and refraction errors of the eye, it is worthwhile to pursue research to test the possible role of sugar in producing these conditions.

18

Does sugar accelerate the life process—and death too?

Sugar's effect on growth

Laboratory animals that are used to test diets are invariably weighed regularly, at least once a week. Almost everyone, therefore, who has looked at the effects of feeding sugar has obtained information about what this does to the rate at which the animals gain (or lose) weight. Sometimes an experiment also measures how much food the animals eat; in this way research workers may be able to show that animals utilize their food with varying efficiency–eating the same amount of different diets, for instance, but gaining less weight with one diet than with another. Sometimes, too, but much less frequently, they not only weigh the animals but actually determine the composition of their bodies. By measuring how much fat and how much lean an animal has in its body, the research workers may find two diets that seem to result in the same gain of weight, but yield a different proportion of fat and lean.

Most workers have reported that sugar-rich diets result in a slower gain in weight in young rats, young chickens, and young pigs. When they measure the amount of food the animals eat, it often turns out that those on the sugar diet gain less weight for each 100 grams of food. And when they look at the composition of the bodies of the animals, they sometimes find more fat and sometimes less.

Here are some examples. Male rats fed for six months from the age of six weeks weighed about 410 grams when we fed them without sugar; with sugar, they weighed only about 380 grams. The effect was more noticeable when the diets were rather low in protein; the rats then reached a weight of 320 grams when they had no sugar but 270 grams with sugar. As I pointed out in the previous chapter, this was because the sugar reduced the body's utilization of the protein in the diet. In an earlier experiment with chickens, some American workers showed that sugar had no effect on the weight when protein was adequate, but did reduce the weight gain when the protein was not quite adequate.

However, the question has been raised whether, in the affluent countries where protein is adequate, the accompanying intake of sugar causes an increase in growth. The most active and enthusiastic proponent of this idea is Dr. Eugen Ziegler of Switzerland. In a number of remarkably detailed and forcefully argued publications he drew attention to statistics from many countries of birth weight, of height and weight of children, and of adult height. According to the information he quotes, these measurements are closely related to the amount of sugar in the diet. Here are some of his examples. The birth weight of babies in Basel, Switzerland, increased from an average of 3.1 kilograms to 3.3 kilograms between the years 1900 and 1960, except during the two world wars, when it decreased. These changes parallel the changes in sugar consumption. In Oslo, the height of girls between 8 and 14 years old increased between 1920 and 1950; for 14-year-old girls the increase was more than four inches. The only interruption to this trend was during the Second World War. Again, these changes in height were parallel to the changes in sugar consumption. Also in Norway, the height of adult men increased by about three quarters of an inch between 1835 and 1870, and by another 1½ inches between 1870 and 1930. The average yearly sugar intake increased from 2¼ pounds in 1835 to eleven pounds in 1875 and to 67 pounds in 1937; current consumption is over 90 pounds, an increase of 40-fold over a period of about 150 years.

So far, I have mentioned only the effect of sugar on the gain of height and weight in children, or of weight in experimental animals. Analysis of the bodies of the experimental animals often shows changes in the amount of fat, as I have said, and also changes in the size and composition of some of the organs. In our experiments with rats we have mostly found a moderate decrease in the amount of body fat; in one experiment, from 35 percent of the dry weight of the animal to 30 percent. On the other hand, some workers have shown an increase in body fat, for example in baboons. This is probably no real contradiction. There is reason to believe that the exact effect of sugar depends on the species of animals you study or even on the particular strain of species such as rats. It also depends on the age when sugar feeding begins, on whether you are studying male or female animals, and on how long the experiment continues.

Sugar's effect on maturity

One of the features of affluent countries is the nutritional state of their babies and young children. No longer is there the incidence of nutritional deficiency such as one used to see: the pinched, starved, rickety children that were common in the larger cities. Instead, there is an appreciable number of fat children, many of them beginning to acquire, even well before they are a year old, the condition that will later turn into years of struggle against fat.

One of the characteristics of these overweight babies and children is that their growth in height is accelerated as well, and they tend to reach maturity early. Although few detailed statistics exist, it is agreed that obesity occurs in bottle-fed babies much more commonly than in the breast-fed. A paper in the British medical journal *Lancet* suggested that this happens because of the early introduction of mixed feeding, especially of cereals. What is overlooked is that a common formula for the bottle-fed baby is a powder consisting of, or largely based on, dried cow's milk to which ordinary sugar is added. It is also usual to add sugar to the cereal feed when it is begun, and indeed quite common to add sugar to other foods as they are introduced, even to egg and minced meat and vegetables. Many of the canned baby foods that are now so commonly used also contain added sugar, and this applies not only to puddings and sweets but also to many savory foods. It is good to see, however, that an increasing number of manufacturers now produce at least some sugar-free baby foods.

All this points to the possible role of sugar in producing childhood obesity. But there is now evidence that sugar may also produce other effects in children. One of the very remarkable changes that has occurred in human physiology during the last century is the reduction in the age when boys and girls reach maturity. Because it is easier to detect maturity in girls than in boys (by the date when menstruation begins), more information exists about girls, but studies do show earlier maturity also in boys.

Briefly, each decade has seen a decrease of some three or four months in the age at which puberty begins. In the past 130 years the age at which Norwegian girls have reached puberty has fallen by almost exactly four years, from an average of 17 years to an average of 13 years. The same trends can be seen in Sweden, England, and the United States. In 1905 the average age of puberty in American girls was 14 years and 3 months; today it is just about 12 years. Incidentally, it is quite wrong to think that puberty occurs early in the tropics; it occurs, in fact, much later than in the better-off countries in temperate climates.

The usual explanation of earlier maturation is that it is caused by better nutrition in the wealthier countries, and by fewer attacks during childhood from infectious and other diseases. But Dr. Ziegler has suggested, with a wealth of statistics, that the main cause is an increase in sugar intake. He believes that earlier sexual maturity is part of the total acceleration of growth that sugar induces. Although he has no experimental evidence, he produces a very plausible explanation in terms of the probable effects of sugar on hormonal secretion. I shall discuss this later in some detail.

In our own experimental work, we have made three observations that support the suggestion that sugar results in early sex maturity. When treating cockerels with sugar diets we have noticed that their combs become red and enlarged earlier than those of cockerels fed diets without sugar. At the end of one of our experiments we found that the testes were distinctly larger in the cockerels fed sugar. With pigs, those receiving sugar were seen to be sexually more active, as shown by their frequent attempts to mount one another in the pen. In rats, sugar produces a distinct increase in the size of the adrenal glands, which, among other functions, produce hormones affecting sex development.

In support of Dr. Ziegler's finding is a report by Dr. O. Schaeffer of Canada. The particular interest of this study is that there has been a large increase in sugar consumption among the Eskimos in the Canadian north. Dr. Schaeffer studied Eskimos in three areas and measured birth weights, as well as the heights and weights of adults and children at various ages. In one of the areas the average annual sugar consumption had increased from 26 pounds to 104 pounds in eight years, in a second area from 83 pounds to 111 pounds in one year, and in a third area from 46 pounds to 61 pounds over five years. Birth weights increased in all of these areas—a small increase with the smallest rise in sugar consumption, and a larger increase, amounting to between half a pound and a pound in one year, in the other areas.

Between 1938 and 1968 the stature of adult men increased by nearly 2 inches and that of women by just over one inch. The height of the children increased much more. Boys and girls aged between 2 and 10 years were 2 inches to 3 inches taller; boys of 11 were 4½ inches taller, and girls of 12 or 13 were as much as 8 inches taller. The latter change was accompanied by a lowering of the age at which there was the rapid weight gain associated with puberty: in 1968 this occurred between the ages of 11½ and 13, while in 1938 it had occurred between the ages of 13½ and 15. The Eskimos appear to show a similar, but perhaps even more rapid, advance of puberty to that which had occurred in Western Europe and America.

The increased growth of children, and especially the earlier development of puberty, is generally assumed to be due to an improvement in nutrition, notably an increase in the intake of protein. This was the explanation given for the considerable increase in the growth of Japanese schoolchildren since the Second World War. In fact, however, while the intake of animal protein doubled, the intake of total protein was only 10 percent more, and there is little evidence that the children measured in 1946 were deficient in protein.

The role of protein is even less likely when you consider that intake among the Eskimos had in fact fallen from over 300 grams a day to just over 100 grams a day during the period that Dr. Schaeffer studied. There was also a substantial fall in the protein intake of Icelanders, one of the groups studied by Dr. Ziegler. On the other hand, in all these three examples—the Japanese, the Eskimos, and the Icelanders—the acceleration of growth was associated with a great rise in sugar intake.

Sugar's effect on longevity

Most of our experiments with animals were carried out for a relatively short time and began with quite young animals, often only a few weeks old. We have had little experience, therefore, in gauging the effects of different diets on the life span of rats, or cockerels, or pigs, or rabbits. We did, however, keep one simple experiment going much longer than usual, beginning with 28 rats one month old. Of these, 14 were given a diet without sugar and 14 a diet with sugar. At the end of two years we had eight rats alive in the starch group and only three alive in the sugar group.

More careful observations have been made by two other groups of research workers. One group in Holland fed some rats with a mixture of foods representing the average Dutch diet, and compared them with other rats that were fed the same mixture but with twice as much sugar. I should add that the amount of sugar in the Dutch diet supplies about 15.5 percent of the calories, slightly less than the 16 percent or so in the average American diet and the 18 percent or so in the British diet.

Of the male rats, those fed the standard diet survived an average of 566 days; those fed extra sugar an average of 486 days. The survival time for female rats was 607 days as against 582 days. If the same proportional reduction in life span occurred in human beings, the extra sugar would result in the biblical "three score years and ten" being reduced to about 60 years for men and to 67 years for women. The greater resistance of the female animals to sugar is another matter I shall discuss later.

The second study on longevity was carried out by some American workers from the U.S. Department of Agriculture. The diets were made up so as to contain either starch or sugar as the carbohydrate component. The investigators studied two strains of rats and, as I have mentioned, found that the strains responded differently to diets containing sugar. One lived just as long with either sugar or starch, although the sugar produced larger livers containing more fat. The other strain also had larger livers with more fat when they were fed sugar. In addition, however, their kidneys were enlarged, and the rats died substantially earlier, at 444 days instead of the 595 days of the starch-fed rats. If you again take the longer survival period as equivalent to 70 years for a human being, the life span with a sugar-rich diet was reduced to the equivalent of 51 years.

There is no evidence at present that sugar affects the life span of human beings. But in the light of this animal research it would not be an entirely absurd suggestion. One keeps hearing how much healthier people are now in the wealthy countries because of improvements in nutrition and the reduction of infectious diseases. As a result, it is reported, the average expectation of life has risen from about 40 years a century ago to over 70 years now. But the former low average expectation of life was due largely to a high mortality in babies and young children; once people reached the age of 25 or so, they were likely to survive to almost the same age as Westerners do now. This is in spite of all the advances in nutrition and medicine and hygiene, so it is reasonable to suppose that these improvements in health have been at least partly offset by some deterioration that holds back what otherwise might have been a slight but very real increase in life span.

That sugar might affect growth, maturation, and longevity is only astonishing if one continues to believe that all dietary carbohydrates have the same metabolic effect once they have been digested and absorbed. It not only ceases to be astonishing but becomes highly plausible when one remembers that sugar can induce sizable alterations in the level of potent hormones.

19

How does sugar produce its effects?

One reason why many people are skeptical about the suggestion that sugar is bad for health is precisely that the number of illnesses in which I feel sugar plays a part is so large. When my colleagues and I say that so many conditions can largely be avoided or improved by avoiding sugar, it looks as if we have joined the panacea-mongers.

Take apple cider vinegar, the food faddists say, or brewers' yeast with yogurt, or wheat germ oil, and you will stay young and healthy forever—well, nearly forever. Avoid sugar, *I* say, and you are less likely to become fat, run into nutritional deficiency, have a heart attack, get diabetes or dental decay or a duodenal ulcer, and perhaps you also reduce your chances of getting gout, dermatitis, and some forms of cancer, and in general increase your life span.

It is difficult, certainly, to imagine that the omission of one single food can produce all these benefits, or that its inclusion in the diet can be responsible, at least in part, for so many disparate diseases. Yet I do not believe that my suggestion is in the least implausible. As I have shown, sugar has a wide range of properties that make it a popular constituent of foods and drinks; it is this versatility that is responsible for its use in so many commodities, and contributes toward today's high intake of sugar.

Because of these very varied properties, it becomes more plausible to imagine that sugar can produce such a large number of varied effects in the body. But research workers are not at all sure of the mechanisms by which every one of the effects can be brought about. Much of what follows, therefore, is inevitably theoretical, but it will, I hope, at least serve the purpose of suggesting some of the lines along which further research can be done.

Sugar can be expected to produce its effects in several different ways. First, it can act locally on the tissues in the mouth or stomach before it is absorbed. Secondly, it can act after it has been digested and absorbed into the bloodstream. Thirdly, it might possibly act by changing the types of microbes that live in the intestines. This could result in a change in the microbial products that appear and get absorbed into the blood, and these in turn might affect the body's metabolism.

The evidence that sugar acts in all these ways varies from near certainty to highly imaginative speculation, but I think all of it is worth looking at. Even the speculative will serve a purpose if it leads to research designed to elucidate some of the remarkable properties of sugar in the body.

Local action

The link between sugar and dental disease

There is wide agreement, as I have already mentioned, about the ways sugar is involved in causing dental decay. The bacteria found in the mouth are stimulated to grow and to produce acid by carbohydrates—by starch and by any sugars that are found in our food. Sucrose, however, is a particularly potent cause of caries for two reasons. First, sucrose is the main ingredient that results in particular foods being sticky and adhering to the teeth; biscuits and toffees are notable examples. This in itself would be conducive to caries production, because the carbohydrate they contain is not washed away; as a result, the acid produced by bacterial action comes into prolonged contact with the tooth surface. But secondly, sucrose, unlike other carbohydrates, has the unique property of being readily built up into a material called dextran, which serves as a most effective raw material for the acid-producing bacterium, *Streptococcus mutans*.

The link between sugar and dyspepsia

The patients whom we treated in the experiment mentioned in Chapter 16 were suffering from a variety of conditions, including hiatus hernia, duodenal ulcer, or severe dyspepsia with or without actual ulceration. There is at present a great deal of discussion about the causes of these conditions. But I think we can imagine a way in which sugar can produce or exacerbate an inflamed mucous membrane in the esophagus or stomach; why a low-sugar diet relieves the symptoms; and perhaps even why sugar can actually produce an ulcer.

If you think about the "natural" human diet, by which I mean the diet before the beginning of agriculture, you will see that the constituents of the food would not be irritating to the stomach. This is because they do not have a high osmotic pressure.

Let me explain what osmotic pressure is. It is a property of a watery solution that is measured by its tendency to absorb more water to itself in particular conditions. If, for example, you put a strong sugar solution on fruit, the fruit will shrink because its moisture is, as it were, sucked out by the sugar. Or if you pour sugar on a cut in your finger, it will hurt as it does when you put salt on it, though not so much because salt has an even higher osmotic pressure than sugar has. This again is because the cells of the skin shrivel through having to give up some of their water.

The osmotic pressure depends on the concentration of particles (molecules or ions) in the solution. If you are dealing with a material like starch, which has very large molecules, then even a strong solution will not have much osmotic pressure because it will contain relatively few molecules. On the other hand, a similar concentration of sugar will have a high osmotic pressure because the molecules are small and so there will be very many more of them.

The pre-Neolithic diet, as I indicated earlier, probably contained a fair amount of protein, a moderate amount of fat, and a little starch and sugar. Both protein and starch have large molecules, and fat doesn't dissolve in water at all. So the osmotic pressure would depend mostly on the small amount of sugar in this diet and the very much smaller amount of other materials with small molecules, such as various salts and vitamins in food. This sort of diet, then, does not irritate such tender tissues as the mucous membrane of the upper part of the digestive tract.

Large amounts of sugar, however, especially if taken in concentrated form on an otherwise empty stomach, will be an irritant. You can actually see the irritation happening if you put a gastroscope into somebody's stomach, which allows you to see the stomach lining. If you now get the subject to swallow a moderately strong sugar solution—the equivalent, say, of four or five lumps in a cup of coffee—you can watch the mucous membrane turn red and angry as the irritant sugar reaches it.

The fact is that sugar in the quantities that are part of the average Western diet, and especially taken as it often is, on an empty stomach, will be a source of repeated irritation on the delicate mucous membranes of the esophagus and the stomach. Irritation of the esophagus is the most likely cause of heartburn. As for the stomach, it is not surprising that a high-sugar diet, even for only two weeks, can result in the production of more acid and much more active gastric juice, as we showed in our experiments. Finally, it is widely held that duodenal ulceration is a result of excessive secretion of gastric juice, so that it is also not difficult to see why sugar might contribute to the cause of this condition.

There is another possible way in which sugar might act on the stomach. As I have shown, sugar affects the adrenal glands, and it is known that some of the hormones produced by this gland increase the production of gastric juice. Sugar would then be producing its effects in the stomach both by a local action and by a general action.

Let me repeat that these suggestions are made simply because they constitute a reasonable explanation of at least some forms of severe indigestion. It remains to be seen whether these are the precise mechanisms by which sugar may contribute to the production of duodenal ulcer, for example. But even if the explanation turns out to be different, there is no doubt of the effectiveness in most patients of the low-carbohydrate diet in the relief of the symptoms of severe and chronic indigestion.

The late Surgeon-Commander T. L. Cleave suggested quite a different mechanism for the cause of peptic ulceration and other diseases of affluence. He believed that all "refined carbohydrate" is equally responsible. Both white flour and refined sugar cause peptic ulceration, he suggested, because they are concentrated—sugar being concentrated from cane or beet, and white flour from the whole wheat. He believed that it is the stripping of its protein that changes innocuous whole flour into ulcer-producing white flour. The idea behind this is that the protein is necessary for the proper neutralization of the gastric acid.

I do not find the theory convincing for three reasons. One is that the difference in the protein content between ordinary brown flour and white flour is very little, about 13.5 percent compared with 13.0 percent; the exact figures will depend on the sample of flour and the precise way it has been milled. But even flour made from whole wheat contains only a little more protein, perhaps 14.5 percent.

Secondly, bread is not the only source of protein, so the neutralization of stomach acid does not depend entirely on bread, either brown or white. Bread contributes about 17 grams of protein a day to the average British diet, where the total intake of protein is about 100 grams. The difference between eating bread from the whole wheat and eating ordinary white bread is something like one gram of protein a day, and rather less if you ate the commoner sorts of brown bread rather than whole wheat bread.

Thirdly, our own experiments have shown that if the amount of starch in the diet, mostly from white bread, is reduced and its place taken by sugar, there is a great change in gastric juice. The effects of bread and of sugar are too different for them to be lumped together as equally dangerous "refined carbohydrate," as we pointed out in Chapter 6.

General action

While we don't know for certain how sugar can produce disease, I do believe that some sort of pattern is beginning to emerge. Now we must put up some reasonable theory based on this pattern, so that further experiments will reveal more of the pattern. Of course we shall have to change our theories if they turn out to be wrong. In trying to understand how sugar can be involved in causing so many diseases and abnormalities, two results of our work have especially impressed me. One is that sugar produces an enlargement of the liver and kidneys of our experimental animals, not only by making all the cells swell up a little, but by actually increasing the number of cells in these organs. In technical terms, sugar produces not only hypertrophy but also hyperplasia.

The second effect that seems to be important is that sugar can produce, at least in some people, an increase in the levels of insulin and estrogen and a more striking increase in the level of adrenal cortical hormone; it also produces an enlargement of the adrenal glands in rats. It should be remembered, too, that these effects are more likely to occur when the blood is repeatedly flooded with high levels of the glucose and fructose produced when the sucrose is digested. This in fact is what happens, partly because—as the advertisement tells you—it is rapidly digested and absorbed, and partly because people so often take sugar in food and drink between meals when there is little else in the stomach to delay absorption.

To begin with, the effects on hormones and on liver and kidney should persuade any reasonable person that sugar is not just an ordinary kind of food. Secondly, its effect in producing raised hormone levels makes it possible to see how sugar can be implicated in such a large number of conditions. It also, I suggest, indicates why people may develop one disease rather than another disease. For the hormones maintain a most intricate interrelationship, both in the amounts circulating in the blood at any one time and in their actions on the body's metabolism. It seems to be always true that an increase in the amount of one hormone results in an increase or a decrease of several of the other hormones.

In a general way, the effect is a tendency to restore the state of the body to what it was before. This occurs because some of the actions of different hormones oppose one another, while some enhance one another. But the likelihood is that, after all the readjustments following the increase of one hormone, some actions of the whole group would still not be in balance.

I would expect that the details of the ways in which these attempted readjustments are made vary from one person to another. Imagine a sudden flood of water into a stream. It eventually forces its way through a weak part of the bank. You now repair this rapidly but you can fetch material only from some other parts of the bank: stones and gravel and mud and sand, a little from several places. When you repaired the breach, you weakened other parts of the bank; only the next flood will tell you which part will now give way. It will depend on so many things, and two streams that seem to be identical will almost certainly behave differently when the stress comes.

Of course, you can pretend that the situation is really much simpler. It is not difficult to imagine that sugar causes diabetes because it makes the insulin-producing cells of the pancreas overwork until they become exhausted. And this may in fact be so for some sorts of diabetes. I say this because there is a growing belief that diabetes is not just one disease, or even the two diseases in the young and in the middle-aged to which I referred earlier. So there may be a complex mechanism by which sugar produces diabetes, or some sorts of diabetes, and not enough is known about the disease to try and unravel the mechanism.

With atherosclerosis, I have worked out the possible mechanism simply for my own benefit, because it gives us ideas of what new experiments we should undertake. This working hypothesis starts with the assumption that the underlying cause of the disease is a high level of insulin. The reasons for this belief are several.

First, many people who have definite atherosclerosis have a high level of insulin in the blood. Secondly, several circumstances increase the risk of coronary disease, and they include cigarette smoking, overweight, peripheral vascular disease, and Type II diabetes. Each of the first three, and often diabetes too, is associated with an increased level of insulin. Thirdly, reduction of excess weight, or increased physical activity, both of which reduce the risk of developing coronary disease, result in a fall in insulin levels. Fourthly, experiments with rats have shown that administration of insulin produces an increased amount of cholesterol in the aorta. Finally, it does look as if some people are much more likely to get coronary thrombosis than other people are, so it would be understandable that only some people react to sugar by a raised level of insulin.

But the most cogent reason for believing that insulin, or perhaps some other hormone, underlies the process that ends as coronary disease is the multiplicity of changes that accompany the disease. As I have said several times, we are looking for the mechanism that produces a condition involving not only a raised level of cholesterol and triglycerides, but also a range of other disturbances: in biochemistry, in platelet behavior, and in a number of other characteristics. Only a disturbance of hormone levels is likely to afford an explanation of such a wide variety of changes.

At the moment it seems that the most likely first change is a rise in insulin level. But at least two other hormones are affected; as I showed, there is great interplay between the activities of the various hormones. It may therefore turn out that the first disturbance is in some hormone other than insulin, and that the rise in insulin level is secondary to this. We do not have nearly enough information yet to decide this question, but I am convinced that further work on hormonal activities is by now the most promising line of research that we should be pursuing.

In discussing the possible role of hormones in producing atherosclerosis, it is wise to remember that the sex hormones certainly play a part; that coronary disease is much more common in men than it is in women, but that the difference diminishes after menopause when there is a diminution in the activity of the female sex hormones; and that there is a particularly close relationship between the hormones made by the sex glands and some of those made by the adrenal glands.

It is not yet possible to begin to describe how atherosclerosis develops; not enough is known about it. Yet it is perhaps worth some speculation. Let me suppose that the first change induced by a diet high in sucrose is a change in the amount of enzymes in body cells, such as the muscle cells. You can imagine that, over many years, a continuation of a high-sugar diet results in a decreased ability of the cells to carry out their normal metabolic processes properly. They now become unable to use properly their ordinary metabolic materials such as glucose, for which they require hormones, especially insulin. As a result, the level of glucose in the blood rises.

In order to overcome this disability in the cells, the pancreas increases the amount of insulin it makes and puts it into the bloodstream. The increased insulin enables the cells now to begin to deal with the glucose and other substances. At this point, the situation may lead to the condition of diabetes, or at least those manifestations of the disease called Type II diabetes. But insulin produces many other actions, and on many cells other than muscle cells, and these one may suppose were not affected by the sucrose in the diet. As far as these other activities are concerned there now exists an excessive amount of insulin. One result would be to change the balance of several of the other hormones. Another result would be to produce effects such as increased fat formation or obesity. And still other results would be to increase the accumulation of cholesterol and other fatty materials in the arteries, perhaps to change the properties of the platelets, and altogether gradually to produce the condition known as atherosclerosis.

Not all of these suggestions are original, although I have to take the responsibility of putting them all down here in what may ultimately turn out to be a quite incorrect sequence. And I would be the first to agree that this is an extremely hypothetical picture. I put it down, nevertheless, for two reasons: first, it indicates a possible role of sucrose in atherosclerosis that is not entirely implausible; secondly, it sets up a hypothesis that can help research workers make decisions about what further experiments should be carried out.

I do not think it is worth pursuing my argument because so much has to be speculation. Let me just say that hormone changes certainly affect the skin, the rate of growth of an animal and its sexual maturity, and that there is growing evidence of the relationship between hormones and some forms of cancer. It is enough for now to say that sugar produces many profound changes in body metabolism. It is therefore quite possible to imagine that it can be concerned in a wide range of diseases, including those such as diabetes and atherosclerosis which in themselves manifest profound disturbances of metabolism.

Microbes in the digestive tract

The third way in which sugar might act is by altering the numbers and proportions of the huge numbers of different microbes that inhabit the intestine. They exist and multiply on the residues of food that have not been absorbed or digested. The sorts of food that have been eaten will determine the kinds and amounts of these materials, which in turn will affect the proportion and numbers of the intestinal microbes.

Unfortunately, medical science is still not very knowledgeable about such details in human beings, although it is certain that changes are produced when sugar replaces starch.

While we do not yet know what effect these may have on the rest of the body there does seem to be something to say about the replacement of part of the milk sugar (lactose) by ordinary sugar (sucrose) for babies. It is known that bottle-fed babies, who often have sucrose added to cow's milk so as to bring the total sugar content nearer to the amount in human milk, tend to have gastroenteritis (diarrhea and vomiting) much more commonly than do breast-fed babies who get only lactose. It has also been shown that the stools of breast-fed babies contain many more harmless lactobacilli than do the stools of bottle-fed babies, and far fewer of the potentially harmful coli bacteria. Again, stools of breast-fed babies tend to kill off added harmful bacteria; those from bottle-fed babies allow them to multiply.

These findings suggest that the intestinal contents can make a baby either more or less susceptible to infection. The research workers attribute this largely to the fact that bottle-fed babies get only part of the sugar as lactose and the rest as sucrose.

It has been suggested that diverticulitis, an uncomfortable disease of the large bowel associated with pain and diarrhea, may be in some way caused by modern diets. One widely accepted suggestion is that it comes from eating food with little residue, especially white bread instead of the more fibrous wholemeal bread. Earlier in this chapter I told you why I don't think one can explain duodenal ulcers and other diseases of Western man in this way. I do think, however, that a possible cause of diverticulitis is the increase in sugar intake at the expense of starch. The different types and numbers of microbes that occur when this dietary change is made could well influence the bowel itself, altering both its activity and its resistance to damage.

Sucrose in the blood

I pointed out that the sugar we eat is digested into glucose and fructose before it is absorbed into the blood. This digestion is usually quite complete except when very large amounts of sugar are consumed; in these circumstances very small amounts of undigested sucrose can get into the bloodstream. As we are beginning to find out, sucrose has several potent actions in living cells, and so it is quite conceivable that these tiny amounts, over a long period of time, can produce damaging effects on the body tissues. This is at present pure hypothesis, but it is a suggestion that future research must pursue.

20

Should sugar be banned?

As this book shows, a great deal of our recent research at Queen Elizabeth College has been concerned with the possible harmful effects of a high consumption of sugar, so that we have increasingly caused unease among many of our industrial friends. Since such a very large proportion of manufactured foods contain sugar, and many of them a great deal, it was to be expected that our relations with one or two friends in industry have occasionally become rather strained.

There have, in fact, been many different reactions from industry, and they were well summarized when I had occasion to meet the four or five directors of a large food manufacturing firm whose wide range of products includes a considerable quantity of chocolate and sugar confectionery. This was several years ago, when the case against sugar was not as strong as it is today, but I nevertheless put this question to them:

Supposing our opinion turns out to be backed by incontrovertible evidence that sugar, and consequently some of your products, contribute significantly to deaths due to coronary disease; would you then continue to make your luscious mouth-watering chocolates?

The range of replies represents the whole range of attitudes I have found among those with whom I have discussed the question of what to do about the high consumption of sugar which now, without doubt, contributes to so much disease and death. At one extreme there was the director who said it was not his job to protect people from themselves; he was not forcing people to eat his products and if they chose to do so at the risk of harming themselves, it was of their own free choice. At the other extreme, a director said that if he were convinced that sugar was dangerous to health he would resign from the company; in the same way, he said, nothing would induce him now to be a director of, or even own shares in, a company that made cigarettes.

Several other views fell between these two extremes. One came from a director who said that if the evidence against sugar became strong he would encourage his firm to put money and effort into research designed to find ways of combating its ill effects—some sort of antidote, for example, that they might put into their products.

My own view? This is based on the belief that I expressed earlier—that people have become increasingly able to separate wants and needs, to an extent that the satisfaction of wants without hindrance can be disastrous for the individual and for the human species. People always wanted to eat sweet foods because they liked them. So long as the only sweet foods they could find were fruit, by satisfying their wants for sweetness they helped to satisfy their needs for vitamin C and other nutrients. But since they began to produce their own foods, and especially since they developed the technology of sugar refining and food manufacture, they have been able to produce and separate sweetness from all nutrients. What people want is no longer necessarily what they need. Because of the strong drives that originally served important biological purposes, it is not enough to say that people should be told what is good for them, and what is bad, and then left to make their own decisions.

In fact, this alleged principle of knowledge coupled with free choice is not as inviolable as is sometimes made out. It is accepted in most countries that people should not have a free choice to smoke opium if they wish, or to sniff cocaine. So the only question is: at what point should the community intervene to protect individuals from following those instincts that our technological skill has made it dangerous to follow?

A continuum stretches from a situation where society should obviously interfere—the smoking of opium, say—to a situation where we cannot effectively interfere—for example, the taking of insufficient exercise. Somewhere in between these two extremes lies the smoking of cigarettes and the consumption of sugar.

It is in this area that most people would agree that efforts should be made to persuade the public to adopt measures that would preserve their health. Sadly, there has been insufficient official appreciation of the need to study seriously the efficacy of the various techniques of persuasion, in the same way as one might study the efficacy of the various techniques of surgery in the cure of disease. This indifference was made clear some years ago when a Member of Parliament asked whether the British Medical Research Council was looking into ways in which people might be influenced to give up smoking. The answer from the government minister concerned was that this was not the proper job of the Medical Research Council. One fears that the same reply would be given today to the question of how to persuade people to stop taking sugar.

One reason why people are reluctant to believe that it is necessary to do anything about studying the art of persuasion is that they do not appreciate the wide gap that exists between knowledge and behavior—between knowing and doing. As I said earlier, it is commonly believed that all you have to do in the way of health education is to inform people. Just tell them that eating sweets makes holes in their teeth, and your job is done. And it is only slowly being realized, even by such United Nations special agencies as the World Health Organization and the Food and Agriculture Organization, that this approach is one of the main causes of the failure of health education in developing countries. It simply is not good enough to tell people that they should eat fruits, or give their babies milk; there is a lot more to it than that.

I have seen many campaigns backed by dental authorities to reduce dental decay in schoolchildren. Sometimes they are content simply when they have produced attractive posters; sometimes they go further and give prizes to children who can answer questions about the structure of teeth and how the process of tooth decay occurs. But rarely have they tested whether their propaganda has in fact resulted in a reduction in the number of teeth developing caries, even though nothing short of this is really of any use. So you can understand why I believe that we should not assume that the danger of eating sugar will be dealt with satisfactorily just by making sure that people are informed; that people will stop taking these foods and drinks once they know that sugar is involved in causing not only overweight and dental decay, but also heart disease, chronic indigestion, ulcers and diabetes, and perhaps a number of other diseases. The likely outcome is, as it has been with cigarette smoking, that some people will be persuaded to stop, but that many will do nothing about it, even if one can convince them of the harm that sugar does.

Should society then in some way coerce people to give up sugar? Most people would answer this question with a very firm "No." It is enough, they believe, that people should be informed about the value of different foods, good or bad, and then left to make their own choice. I have given my own reasons why I think that our ability to separate palatability from nutritional value makes this an unrealistic view. Moreover, the idea that free choice is sufficient implies that the choice is in fact free; that people do have total and unbiased access to knowledge about food values. But do they?

Those who like myself are worried about excessive consumption of sugar—dentists, for example—often point to the enormous volume of advertising for confectionery, cakes, ice cream, and soft drinks. In Britain alone, more than 100 million pounds a year is spent in advertising these goods. But I am not sure that advertising does very much to increase the total amount consumed. There is some evidence that the effect of advertising is, rather, to persuade people to buy one brand instead of another brand—Coca-Cola instead of Pepsi-Cola, say.

I am not convinced that the media's policy on the acceptance of advertising works entirely for the benefit of the consumer; I feel they tend to look over their shoulders just a little nervously to make sure that they have not offended the advertisers or their agents. And I am frankly very skeptical when I read the claims of the British and American advertising industry that they always have the interests of the community at heart. The Chairman of the British Advertising

Association has said that its objectives include "keeping the pathway open for honest advertising—paving it with honesty, widening it with new understanding, getting it recognized as a utility serving the community as a whole." I am sure everyone can think of examples of advertising that are far from serving these objectives.

With many examples in mind of how information can be distorted or withheld, it becomes even more evident that people should not be left entirely to themselves to decide what they should or should not eat. Sooner or later, I feel, it will be necessary to introduce legislation that by some means or other prevents people from consuming so much sugar, and especially prevents parents, relatives, and friends from ruining the health of babies and children.

But so long as this is not considered a public health matter, is there nothing we ourselves can do? Some people find it quite easy to give up sugar, but many find it really difficult. Let me tell you how I managed. I must now confess that I used to be about the most dedicated sugar "addict" that you have ever seen. I stress this for two reasons. One is that a lot of people imagine that my campaign against sugar comes about just because I don't like sweet things; if only they knew how many pounds of milk chocolate and licorice allsorts and cakes I used to tuck away each week! At a rough guess, I would say that my total sugar consumption must have been not less than 10 ounces a day, probably nearer 15. The second reason for this confession is to show that it is possible to break the sugar habit. I have cut down from five or six pounds a week to at most two or three ounces a week—sometimes next to nothing—and if I can do it, so can you.

The first thing of course is to have the incentive. You must make up your mind quite firmly that you really want to reduce your sugar intake. It may be that you are beginning to worry about your waistline, or your dentist's bills, even if you don't really believe all I have said about ulcers and diabetes and heart disease. Once you have made up your mind, then you won't find it too difficult. But start slowly. If you take two spoons or lumps of sugar in your coffee or tea, cut it down to one for a week or two, and then to a half for a week or two, and only then stop altogether. Try not to drink the usual soft drinks. Drink low-calorie drinks instead, or iced tea; and what is wrong with plain water? If you really cannot drink less beer or cider, choose the dry varieties. And avoid the ordinary "mixers" for your whisky or gin or vodka.

You can also cut down gradually on puddings and ice cream, and you can look out for the less sweet varieties of cakes and biscuits. Keep off the sugar-coated cereals for breakfast, and of course don't sprinkle sugar on them.

You may find it difficult to believe, but when you really have got used to taking very little sugar in your foods and drinks, you will notice that all your foods have a wide range of interesting flavors that you had forgotten. Swamping everything with sugar tends to hide these flavors, and blunts the sensitivity of your palate. You will especially notice how much you enjoy fruit, all the subtle differences between one sort of apple or pear or orange and another. And unless you eat a couple of pounds or more of fresh fruit a day, you can't possibly get to take in as much sugar as the average person now eats of refined sugar, let alone the even greater amount that so many people eat.

All this does *not* mean that you must never, in any circumstances, take a piece of pie or a helping of ice cream. No great harm will come to you if, at a dinner party, you accept something special that your hostess has made for the occasion. Eating sensibly is not the same as making a nuisance of yourself. There are clearly some sources of sugar that are likely to give you much more than do other sources. If you find that you usually put two or three pieces of sugar in your tea and coffee, and if when you add up you find you are taking seven or eight cups a day, you can easily see that you have here a chance of reducing your sugar by two or three ounces a day. Add the amount you take with your breakfast cereal, and perhaps in the occasional cola or fruit drink during the day, and you will find that it is not a great hardship to get down to a quarter of your usual intake, or even much less.

It is more than likely that the harmful effects of sugar are greater when you take it with little else. Eaten in this way, its digestion and absorption are not hampered by the digestion and absorption of other foods, so that the bloodstream is quickly flooded with sugar.

So it is more important to avoid sugar taken between meals, for example in drinks and confectionery, than, say, a piece of apple pie taken at the end of the meal, when the digestion and absorption of the sugar will be very much slower, and its effects much less.

Perhaps the most difficult problem is how to bring up your children without smothering them with sugar. Everything in our modern way of living seems to conspire to thrust sugar down their poor innocent and uncomplaining throats, almost from the moment they are born. But with a little care you can at least see that your children do not get into the "two or three pounds of sugar a week" bracket.

You should begin by choosing one of the baby formulas that is made up with added milk sugar (lactose) instead of with ordinary sugar. Next, when you introduce cereals or more extensive mixed feeding, choose instant or canned foods whose labels say, "No added sugar," or take the trouble to make your own sieved meats and vegetables. Make sure the orange juice has had no sugar added to it, or again make up your own.

Later, by all means give the occasional sweet or biscuit, but only occasionally and as a treat. Never, of course, give it at bedtime after your children have cleaned their teeth. A good plan is to get your little ones to clean their teeth after every occasion when they have eaten a sweet or biscuit. Ask them when they come home from school or from a visit to grandma if they have had any sweets, and if so get them to clean their teeth straight away. With luck, they may get bored with so much tooth cleaning and be contented with sweets only at mealtimes, after which you no doubt want them to brush their teeth in any case.

In the end, the difficulties are not so much to do with how you bring up your children but with how much your kind friends and relatives press sweets into their little hands, often behind your back. Although you may not be able to keep them away from sugar as much as you wish, you will find it quite possible to keep the amount down to far less than many children now have.

You will have noticed, by the way, that I prefer the low-calorie soft drinks to those that contain sugar. You will see from this that I do not at all accept that you run any risk from taking the artificial sweeteners that they contain. My own view is that it is highly unlikely that these do anybody any harm, whereas there is no doubt whatever that sugar can do a very great deal of harm. You may of course decide that it is better to wean yourself entirely from taking sweet foods and drinks, and that you can do this more readily by avoiding the use of sugar substitutes altogether. This is a decision you must make yourself; all that matters is that you should take as little sugar as you can.

Before you begin to reduce your sugar intake, and again at the end of your first week, make a list of all the sugar you have taken on an average day. Make a rough calculation on the basis of this table, and see how much you have saved since you began. In particular, see if you have got down to less than 50 grams a day (nearly two ounces) during your first week, and then how long it takes you to get down to 20 grams a day.

Sugar content in grams of some foods and drinks

1 piece of sugar	**4**
1 flat teaspoon of sugar	**5**
1 bottle of cola	**12**
1 glass of "fruit drink"	**20**
1 spoon jam or marmalade	**5**
1 2-oz. piece of cake	**10**
1 4-oz. piece of apple pie	**20**
1 2-oz. piece of chocolate	**30**
1 oz. sweets	**20**
1 2-oz. ice cream	**12**
1 oz. cornflakes	**2**
1 oz. All-Bran	**5**
1 oz. tomato ketchup	**5**
1 oz. chutney	**12**
1 oz. sweet pickle	**5**
1 oz. salad cream	**3**

It is true that very many other manufactured foods have had sugar added to them; some of them are mentioned on p. 53. But a look at the label will tell you whether it is likely to be a large or a small part of the product, and you can then work out whether the amount that you will be taking of the pickle or the soup or the meat stew is likely to add much to the total amount of sugar in your diet.

21

Attack is the best defense

One way in which the sugar industry responds to attack is to try to put pressure on the other food industries that seem to be drawing attention to the harmful effects of sugar. An example is a talk I once gave that was sponsored by one of the large international food manufacturers. It was published in a book, together with several other talks on nutrition by other research workers. In my talk, I again had occasion to refer to research on the undesirable qualities of sugar. Soon after the book was produced, the chairman of the food company that had organized the talks and was distributing the book was approached by the chairman of a sugar refining company, and asked to stop the distribution of the book because it was not seemly for one food manufacturer to "knock" the product of another. After some argument, the book's distributor agreed to do this; the sugar man was not to know that only two out of the several thousand copies had not yet been sent out.

An obvious way to respond to attack is simply to deny its basis; an even more subtle way is to claim that exactly the opposite is true. If most people say that sugar causes dental decay, you must keep on publishing advertisements or short articles in which you stress that sugar is not important; what is important is constitutional proneness to dental decay, or whether one uses the toothbrush often enough. And when most people say that sugar makes you fat, you mount a campaign in which you claim that in fact sugar makes you slim. We saw some examples of this earlier.

The most intensive publicity activity of the sugar industry has been its attack on cyclamate. This campaign was pursued even though, as I showed, sugar interests like to claim immunity from attack by other food producers.

On the other hand, the sugar industry has supported very little research as to what sugar does in the body. It did, it is true, for several years support research on sugar and dental caries, but even some of this support has been withdrawn. I myself have several times invited the International Sugar Research Foundation to support the work we were doing in my laboratory, on the grounds that the sugar people themselves ought to be the first to know whether their product does in fact produce ill effects. Two or three times it really appeared that they were going to help us financially in our research, but each time the suggestion fell through.

The International Sugar Research Foundation has, on very rare occasions, supported experimental work directly relating to the possible involvement of sugar in producing disease. For example, a research report appeared in the middle of 1971 from Wake Forest University in North Carolina. A dozen miniature pigs were fed on diets with sugar and compared with a dozen fed without sugar. Six pigs in each group were killed at the end of one year; the remaining six in each group were killed at the end of two years. The International Sugar Research Foundation has triumphantly claimed that the results prove that sugar does nothing either to the cholesterol level or to the development of atherosclerosis. A careful look at the results, however, shows that the cholesterol in the sugar group was, as it happened, somewhat lower than that of the control group at the beginning of the experiment; thereafter it was almost continuously higher. Moreover, there was in fact more atherosclerosis in the sugar-fed pigs than in the control pigs.

Can you wonder that one sometimes becomes quite despondent about whether it is worthwhile trying to do scientific research in matters of health? The results may be of great importance in helping people to avoid disease, but you then find that they are being misled by propaganda designed to promote commercial interests in a way that you thought only existed in bad B films.

Some of my best friends . . .

Every so often people are told that they should eat, or should not eat, a particular food because of its effect on their health. The publicity may or may not be well founded; what is important to the producer or manufacturer of the food is whether it is believed and acted on. If people really believe that they are less likely to suffer a heart attack if they eat margarine rather than butter, the margarine manufacturer will rejoice and the butter producers will be saddened. And, understandably, both will take steps to advance or protect their commercial interests.

It is then not unreasonable when the producers and the refiners of sugar, and the manufacturers of sugar-rich products, react vigorously to publicity suggesting that sugar is harmful to health and its consumption should be curtailed. What may be considered less reasonable are some of the particular ways these organizations react. A few of these have come my way, and in this chapter I give some examples that may be interesting to those who wonder whether there is justification for the concern expressed from time to time about the power that is in the hands of the "multinationals."

The World Sugar Research Organization, or, What's in a name?

In the year or two after the UK publication of *Pure, White, and Deadly*, the book was translated into Finnish, German, Hungarian, Italian, Japanese, and Swedish. By 1979 it clearly needed updating, since there had been quite a number of new discoveries about the effects of sugar. Although the publishers were pressing me to produce a new edition, I was then too occupied with other activities to have the time for what would have to be a fairly extensively rewritten book. So the English edition went out of print.

This fact was not overlooked by the sugar industry. The *Quarterly Bulletin* of the World Sugar Research Organization (WRSO), published from the London headquarters, is a sort of newsletter containing mostly summaries of research that bring good tidings to the industry. On the whole, these are from articles that either comment favorably on the use of sugar, its production or marketing, or that draw attention to some unfavorable aspect of the use of sugar and are then criticized in the *Bulletin*.

In 1979 it published the following under the headline, "For your dustbin":

"Pure, White and Deadly." J. Yudkin. Davis-Poynter Ltd, London 1972.
Readers of science fiction will no doubt be distressed to learn that according to the publishers the above work is out of print and no longer obtainable.

Like any serious research worker, I do not mind people disagreeing with whatever conclusions I draw from research—my own or that of other serious research workers. But to say that my work is "science fiction" is to say that what I had published as representing the results of my research and that of my departmental colleagues, as well as the research by other scientists I had quoted, was invented and imaginary.

My view of the statement published in the *Bulletin* was shared by all those colleagues who saw it. My solicitor, who had had great experience in libel cases, was of the same opinion, but wisely sought the opinion of two separate barristers, both specialists in libel law. They also took the view that it is libelous to suggest that a scientist whose work has been published in British and foreign scientific journals of repute has in fact been presenting fictitious research findings.

We initiated an action for libel, which began a four-year exchange of letters between lawyers. In the end, the sugar organization and its editors agreed to publish a retraction, and to pay my legal costs, which up to that time had not reached too high a level. We therefore settled with the organization and abandoned the suit. Here is the statement that was published in the *Bulletin* in March 1984:

In the Quarterly Bulletin of September 1979 we commented on the fact that the book, "Pure, White and Deadly" by Professor John Yudkin had gone out of print. We also made other comments relating to the contents and value of the book. We are sorry that the publication of those comments has been taken by Professor Yudkin to impugn his integrity or reputation as a scientist.

Professor Yudkin is internationally known for his work on nutrition, having written a large number of research papers that have been published in a wide range of scientific and medical journals of the highest repute. He is also the author of several widely read books on nutrition, a subject with which his studies have been principally concerned. He has over the years acted as a consultant to a number of companies concerned with the manufacture of food or ingredients relating to food, including Ranks Hovis McDougall, Unilever and the National Dairy Council. Based on a series of experiments which he has been carrying out since the late 1950s he has formed views for which he is well known to the effect that sugar is not a safe commodity for human consumption. We accept that he holds these views and no imputation is cast upon his sincerity or the good faith of his research. Professor Yudkin recognizes that we do not agree with these views and accepts that we are entitled to express our disagreement.

An ironical aspect of this affair was that the then Editor of the *Bulletin* was at the time a member of the Council, that is, the governing body, of Queen Elizabeth College, where I had been Professor of Nutrition for many years. He had been appointed Honorary Treasurer, and had been a member of the College Council in 1976, when, five years after I had formally retired, it had elected me a Fellow of the College—an honor that had otherwise been given only to retired administrative members of the College. He must therefore have voted for, or at least acquiesced in, my election as Fellow, which took place "in recognition of [my] contribution to the reputation of the College in helping to establish and build a flourishing and highly respected Department of Nutrition."

During the prolonged period when the lawyers were exchanging letters about my "work of fiction," I attended an informal party at the College, where I was buttonholed by the Principal. He took me aside and told me that he had heard I was seeking to sue the Treasurer of the College Council. Just as I was about to thank him for sympathizing with me for being maligned by the Treasurer, he made clear his view that it was I that was at fault for attacking an officer of the Council of my own College. I thought that it would have been more appropriate if he had suggested that the Treasurer should resign from the Council for his unwarranted attack on an Emeritus Professor of the University and a Fellow of the College.

Freedom of choice depends on freedom of information

By itself, the affair I have just described would be of little public concern. But it is only one small example of the activities of the various organizations that comprise the multinational sugar industry.

Take smoking. When there is talk of helping to prevent lung cancer or chronic bronchitis caused by smoking by controlling advertising or by increasing taxes on tobacco, there is considerable protest, mostly from the tobacco industry, that such actions curtail freedom of choice. We are told that society has no right to interfere if a person is prepared to take the risk of dying of cancer, or of being unemployable because of severe bronchitis. But freedom of choice exists only if there is freedom of information. The sugar industry has constantly attempted to prevent the public from being informed about the harmful effects of sugar. To substantiate this accusation, let me cite some of my own experiences during the past 20 years or so.

Early in 1964 I received an invitation to read a paper on the research we had been carrying out on food habits. The invitation came from the secretary of an organization in Paris called La Fondation

Internationale pour le Progrès de l'Alimentation (FIPAL). I was told that the organization was supported by the food industry, but that its work was uninfluenced by commercial considerations. In July of that year, I published in the *Lancet* some of our findings, including evidence suggesting that sugar was a cause of coronary heart disease. Shortly afterward I received an agitated letter from the secretary of FIPAL, asking whether there was any truth in the reports of this research that had appeared in French newspapers. The reason for this letter, the writer said, was that, as well as being secretary of FIPAL, he was also secretary to the French body concerned with promoting sugar.

My reply to this was that the reports of our work were correct, and I suggested that in the circumstances it might be better if I withdrew from the conference. This elicited a strong denial that his letter implied any suggestion that my presence at the proposed meeting was not welcome. The secretary repeated the statement in his first letter that FIPAL's sole objective was to promote work and discussion on nutritional problems.

The meeting took place in September that year, and papers were read by some dozen research workers. One of my colleagues accompanied me to Paris, and he and I were asked to stay for two or three days after the meeting in order to edit the contributions for publication. Some months later I was sent the proofs of my own paper, with a request from the secretary: since I had mentioned that there was now evidence that the recent considerable increase in sugar consumption was a possible cause of the increase in some diseases, would I please withdraw this statement or put in a footnote that this was a personal opinion that was not universally agreed? I wrote to say that this request was not compatible with his early assurances of the impartiality of FIPAL; I suggested that if they did not wish to publish my paper as I had read it, I would rather they did not publish it at all.

That is precisely what happened. The book appeared with my name in the list of those who had contributed to the meeting, but you will find no record of what I said there.

Sugar and artificial sweeteners

You would expect the sugar industry to keep up a steady campaign against the use of artificial sweeteners like saccharin and cyclamate, and some of the newer products like aspartame. This campaign is now much less active than it used to be, since the sugar refiners are themselves in the process of developing new artificial sweeteners. Nevertheless it is still interesting to look at some of their earlier activities in this field.

Take the cyclamate affair. The sugar industry spent a great deal of money on research and publicity on the possibly harmful effects of cyclamate. They announced this repeatedly in their information reports right up to 1969, when cyclamate was banned in the USA, the UK, and some other countries. Here is a quotation from an American sugar agency as early as 1954, explaining why sugar was spending so much money on publicity:

"These substitutes might never command a really damaging share of the market in terms of bottles, cans, and cases, but their share of market in terms of human prejudice might be very damaging indeed. This obviously calls for a broad programme of information about sugar among consumers. It is the only real insurance the industry can have."

By 1964 the sugar industry had come to the conclusion that artificial sweeteners really were a serious challenge. The President of Sugar Information Incorporated, addressing the Sugar Club, then said, "Every man in this room is affected directly in the pocket-book, by the challenge of the synthetic sweeteners. I want to discuss with you the nature of this challenge, its dimensions and its impact. I want to tell you what we are doing to meet it." He then went on to describe an advertising campaign "questioning the value of synthetic sweetened soft drinks."

Some of the experiments with cyclamate that the sugar people sponsored were really not very well carried out. For example, in one experiment rats were fed with a diet containing 5 percent of cyclamate, which is equivalent in sweetening power to sugar amounting to one and a half times as much as the total amount of food normally eaten! Nobody will therefore be surprised that the rats did not thrive on this diet, and did not grow as well as did rats without cyclamate.

But the great scientific discovery about what 5 percent of cyclamate in the diet does to the growth of rats was very widely publicized, not only in articles in many magazines, but in an information brochure sent to, among others, every Member of Parliament in Britain.

The chief irony of the cyclamate story is that the eventual banning of this sweetener in the United States was the result of research sponsored by Abbott Laboratories, the world's largest manufacturer of cyclamate. In this research, carried out by the Food and Drug Research Laboratories in New York, rats were given enormous doses of cyclamate with saccharin, equivalent in sweetening power to 11 pounds of sugar a day. At the end of two years, a very long time in the life of a rat, a few animals showed the beginnings of cancer of the bladder. Ordinarily, one would now get a group of experts together to try and evaluate the relevance of these studies to the human consumption of something like one fiftieth of the equivalent dose—about the maximum amount of cyclamate anyone would take. Indeed, it is now widely accepted that the occurrence of cancer in that experiment had nothing to do with cyclamate or saccharin.

Nevertheless the decision to ban cyclamate was inevitable because of the Delaney Clause in the American food and drug legislation. This, as you recall, says that any material that, in any dose, over however long a time, causes cancer in any animal, must not be used in human food. So cyclamate was banned in the USA, and then in several other countries, thereby presumably inviting everybody to go on eating all the sugar they wanted. Now, however, most countries, having reconsidered the position, have removed the ban.

There is also a personal twist to these experiments and results. When I have reported some of the experiments that my colleagues and I have done, sometimes with as much as 15 ounces of sugar a day in young men, but often with a lot less, I am told that these are abnormally large amounts and that our results are not valid. In fact they are nowhere near the equivalent of the astronomical amounts of cyclamate that had to be used in order to show how "dangerous" this material is.

The British Nutrition Foundation

The British Nutrition Foundation was born in 1967, 26 years after the birth of the Nutrition Foundation of the United States. The latter is funded almost entirely by the American food industry and has a large Council, comprising not only members of the industry but also research workers in nutrition and food sciences and distinguished members of the public. It produces a regular monthly journal, *Nutrition Reviews*, which discusses and comments on recently published research in the wide field of nutrition. The Nutrition Foundation also publishes occasional volumes that summarize what research has discovered in the major areas of nutrition. On the whole, it can be said that the American Nutrition Foundation is not influenced by the fact that it is funded by the food industry, although it has to be admitted that it rarely criticizes aspects of the industry that a completely uncommitted group might consider deserve at least some degree of criticism.

Thus, when it was set up in 1967, the British Nutrition Foundation (BNF) had the American organization as a model, and it too was funded by the food industry. Its first and major sponsors were the sugar refiners Tate & Lyle, and the flour millers then known as Rank. This combination occurred, it seems, chiefly because of the personal friendship between, on the one hand, the families of Tate and Lyle, and on the other hand the Rank family. There was also a business friendship between the two groups, since Rank was to a sizable extent a user of sugar—for example, in the manufacture of cakes and biscuits—especially after it had amalgamated with two other large firms in the flour milling and baking industry to form Ranks Hovis McDougall.

The first Director of the British Nutrition Foundation was the late Professor Alastair Frazer, a biochemist who had taken a special interest in the biochemistry of drugs and had just retired from the Chair of Pharmacology at Birmingham University. His major research had been on how the body digests and absorbs fat from food. To this extent, then, he was concerned with nutrition, although in a fairly narrow field. At first he was kept busy approaching other firms in the food industry, most of whom were, it seems, less than enthusiastic about promising financial support to the organization; for this reason the BNF had a precarious first few years. However, one approach to the food industry seemed more successful than most: Professor Frazer's claim that, in a climate of growing consumer concern about processes and additives used by the food industry, the BNF would stand as a sort of protective fence between the industry and the public. In spite of these time-consuming efforts to produce funds for the Foundation, the Director-General nevertheless found time to supervise and support a film telling of the virtues of sugar as a food.

From what I have said, you might ask whether the BNF at that time tended to be somewhat on the side of sugar, and if so whether it has remained so. I shall let you make up your own mind when you have finished reading this chapter.

The Director-General objects

In the late 1960s Ranks Hovis McDougall (RHM) decided to begin research into the possibility of producing an inexpensive high-protein food: an attempt that, some twenty years and tens of millions of pounds later, has recently resulted in an excellent savory pie appearing on the market. At the very beginning of the project I was asked by the then Director of Research of RHM to act as an adviser on the project.

At the same time he told me that his friends from Ranks Hovis McDougall and from Tate & Lyle, both of which continued to be major sponsors of the BNF, had said that it was not appropriate for me to advise RHM; nevertheless he himself wanted me to do so. A short while after the project got under way, he told me that the Director-General of the BNF was pressing him to tell me to desist from saying that sugar was harmful. I said that it would be more sensible if we had a meeting with Professor Frazer at which I would describe the results of our recent research and explain the reasonableness of my views.

We met at BNF headquarters: Professor Frazer, the Research Director of RHM, two or three members of BNF, and I. We had an interesting discussion, from which it was clear that Professor Frazer was not very up to date on research into the causes of coronary heart disease, or research into some of the effects of sugar on the body. He strongly rejected the suggestion that sugar had, or could have, anything to do with coronary disease. He insisted that there was no relationship between the increase in sugar consumption and any increase in coronary disease; in fact, he said, there had not been an increase in the disease. I said that this was contradicted by the general recognition of cigarette smoking as an important cause of the disease; as there had been a tremendous increase in smoking, it followed there must also have been an increase in the prevalence of heart disease. "That only shows," said Professor Frazer, "that smoking too has nothing to do with the disease"—a view that would have been supported by very few other scientists or doctors.

As we left the room after lunch, the Director-General was overheard to say, "You can take it that Yudkin won't be getting any research grants from the BNF"; this prophecy was certainly fulfilled.

The BNF doesn't want nutritionists from QEC

Throughout my time as Head of the Department of Nutrition at Queen Elizabeth College, neither I nor any of my colleagues had any association with the BNF. I should point out here that my Department, instituted in 1953, was the first in any European university to be devoted to undergraduate and postgraduate teaching of nutrition, and was carrying out research that was probably at least as extensive as that of any other nutrition department in the country.

In terms of the aims of the BNF, its most important committee must be its Science Committee. The chairmen of this committee have always been distinguished scientists; none has been a professional nutritionist but they have all had some contact, if sometimes rather remote, with the subject of nutrition. As I write, there have been five chairmen of this committee since the Foundation began; these have included the late Sir Charles Dodds, one of the outstanding biochemists of the time, and the late Sir Ernst Chain, who shared the Nobel Prize for the discovery of penicillin with Florey and Fleming. Both Dodds and Chain approached me while Chairman and asked why I was not on the BNF Science Committee, or indeed on any of its other committees. When I said that I had not been invited, they asked if they might suggest that I should be appointed. To this I agreed, although I guessed what the reply would be. And so it proved. Both chairmen had been told in due course that there was no question of having me in any way associated with the BNF. What I had not guessed was that the member of the BNF Board from Tate & Lyle, which had remained one of the major sponsors of the Foundation, had said that if I were appointed he would resign from the Board, and would see that his firm—and others—withdrew their sponsorship.

After it was founded in 1953, the Nutrition Department of Queen Elizabeth College rapidly became a thriving center of nutrition research, and was soon responsible for having trained several of the graduates doing nutrition research in other laboratories in this country and abroad. We were clearly interested, therefore, when in 1970 it was announced that a joint committee of the Agricultural Research Council and the Medical Research Council (ARC–MRC Committee) was being set up to examine the current state of nutrition research in the UK, and what important problems most needed investigating. To our surprise, neither I nor any of my staff were appointed to the ARC–MRC committee.

After the report had been published I happened to be writing to the Chairman of the Committee, who was a long-standing friend. In the course of my letter I said that it would interest me to know why no one from my department had been invited to join his committee, in view of our position as an important nutrition research center. He replied that, since he himself was not a nutritionist, he had taken advice from people in the field. He had consulted the British Nutrition Foundation, and it was they who had told him that I was not an appropriate person to be on the Nutrition Research Committee.

The long arm of the sugar industry

You may well consider that my experiences with the British Nutrition Foundation reflect a rather remote and perhaps unimportant sort of intervention of sugar interests in the affairs of academic workers carrying out research and disseminating its results. Let me then mention two rather more direct interventions.

Those of you who have been to Switzerland will no doubt have seen one of the many elegant branches of the supermarket chain Migros, or will have bought petrol in one of the Migros garages. During his lifetime the founder of this large organization, Gottlieb-Duttweiler, set up a trust whose income is a percentage of the turnover of the business. Among many other activities, it organizes occasional symposia on subjects of international concern, such as ecology and nuclear energy. In 1977 the Gottlieb-Duttweiler Institute appointed Al Imfeld to organize these symposia, beginning with one that was to consider the subject of sugar—its production and distribution, its political and economic background and activities, and its role in human nutrition. Al Imfeld asked me to be one of the speakers at this symposium and invited me to read a paper on the nutritional role of sugar. Soon after I had sent him my proposed paper, and a month or two before the meeting was due to take place, Imfeld wrote to say that the meeting had been canceled and that he had been dismissed from his job; he added that he knew that I would understand the reasons for these events.

The Gottlieb-Duttweiler Institute did hold a meeting on sugar in 1981, although this time I did not receive an invitation to attend.

It was a somewhat bowdlerized meeting in that none of the speakers dealt with the international financial and political activities of the sugar companies, as had been intended at the meeting planned originally by Imfeld. Nevertheless, it was interesting to read in the report of the meeting what was said by Eugenie Hollinger, the representative for consumer affairs of the Migros organization: "I well remember the appearance of the German translation of John Yudkin's sugar report, *Süss aber gefährlich (Pure, White, and Deadly)* in 1974. I had the greatest difficulty at that time in persuading any newspaper publisher that the book should be reviewed. They were all afraid of an advertising boycott by the affected food industry and distributors."

Subsequently Imfeld published a book with the simple title *Zucker*, which strongly indicts the worldwide activities of the sugar industry and explicitly points out the role it played in bringing about the abandonment of the Institute's original meeting and the loss of his job.

My second example occurred three or four years later. A new artificial sweetener, aspartame, was about to be given government approval in the UK, USA, and several other countries. Aspartame is produced by the American pharmaceutical company G. D. Searle, which has a large operation in England. I was approached by the English company to organize a conference dealing generally with carbohydrates in nutrition, although there would be in addition one speaker from Searle who would give a paper about the new and still little-known aspartame. I spent a great deal of time corresponding with possible speakers, from the UK and from other countries, and discussing the particular areas that they would be asked to cover. Bookings were made for the travel and accommodation of the participants, as well as arrangements for the conference itself at a large hotel in Stratford-upon-Avon. Two weeks or so before the meeting was to take place, it was canceled. At this late stage I was left with the unpleasant task of informing the speakers with whom I had been carrying on prolonged and detailed correspondence, and who by now had prepared the papers they had intended to give at the conference. More difficult still, I had to try as tactfully as I could to avoid telling them what I understood to be the real reason for the cancellation.

The person from Searle who for months had been making the manifold technical arrangements for the conference told me the news about the cancellation; he was understandably very upset and angry. It was therefore not surprising that he could not restrain himself sufficiently to maintain the secrecy that the company presumably intended concerning the reason for the abandonment of the conference. According to him, it was the Coca-Cola Company that had pressed Searle to cancel the meeting. Coca-Cola are the world's largest single users of sugar. In 1977, I am told, they used one million tons of sugar in the USA, so they had a considerable interest in what the public were told about sugar. Meanwhile they were also producing Diet-Cola for people wanting low-calorie soft drinks; although making up only a small proportion of total soft-drink consumption, this was nevertheless a large and thriving market. Thus, in the early 1980s, Coca-Cola was negotiating with Searle about using aspartame in these drinks instead of only saccharin—an enormous potential market for the new sweetener. This fact gave Coca-Cola the opportunity to suggest that their decision could depend on whether Searle proceeded with the conference, which would undoubtedly have publicized new research on the ill effects produced by the consumption of sugar. And Searle abandoned the conference.

Telling the truth about tooth decay

The most impressive campaign to inform people of the ill effects of sugar consumption was, I believe, the one begun in 1977 by the North-Rhine Dental Insurance Association (the Kassen-Zahnärtzlichen Vereinigung Nordrhein, or KZV). This was done mostly through the enthusiastic and energetic activities of its chairman, Dr. Edvard Knellecken. With more than £1 million a year that they put aside for anti-sugar propaganda, KZV advertised in newspapers and magazines; wrote letters to doctors, scientists, and politicians; and campaigned for a range of legislative measures to combat the promotional activities of the sugar industry. They suggested that packets of chocolates and confectionery should have printed on them some symbol such as a toothbrush to indicate the potential damage to the teeth from the consumption of these products. They asked that no suggestion should be allowed in advertising that sugar promoted health or fitness, or performance at sports. They asked for a tax to be put on sugar itself, and on all sugar-rich food and drinks, as there is on tobacco and alcohol.

The KZV called a widely publicized conference at which the media were well represented and where speaker after speaker described the ill effects of sugar consumption and the research that had been done to demonstrate this. I gladly accepted their invitation to this conference and was the only non-German present. I spoke of our research on sugar in relation especially to heart disease and diabetes.

We were not surprised that the publicity achieved by this meeting was followed by strong reaction from the various branches of the sugar industry. One of these was of special interest to me; it was a copy of a letter received by Dr. Knellecken, written by an Austrian doctor, Dr. Göttinger. Here is a translation of part of the letter:

> Thank you very much for kindly sending me your information about dental caries.
>
> I am surprised that it seems to have escaped your attention that dental caries has for a long time been accepted as an infection, and vaccines against the condition are already being prepared . . .
>
> Perhaps it has also escaped your attention that Professor Yudkin is not a university professor, and has no professorial chair. He is, rather, a grammar-school teacher in London, as stated in his books, and he has also never carried out any experimental work, but used only statistical arguments. I know his books and have some of them. In the opinion of many authoritative people, he is in fact not a scientist to be taken seriously.

I have wondered what motive Dr. Göttinger had in making such an outrageous and unwarranted attack on a medical colleague. I wrote to him correcting his misapprehension by pointing out that I had a string of university qualifications, was holder of the Chair in Nutrition and Dietetics at London University, and had published getting on for 300 research articles in many scientific and medical journals of international repute, as well as several books, which it was clear he had not read. You will perhaps not be surprised to learn that Dr. Göttinger did not reply to this or to subsequent letters; I do, however, still get from him a request for a reprint each time I publish a new research paper.

Sadly, the activities of the KZV were interrupted when Dr. Knellecken was accused of financial fraud in relation to the funds of the association—accusations that were instigated by the sugar industry. As a result, KZV's attempt to inform the German people of the considerable damage that sugar does to their health was brought to a sudden standstill. Some three or four years later, however, I was delighted to see that, although belatedly, Dr. Knellecken's reputation had been wholly vindicated, as was made clear by a report in the German magazine *Naturartz*. This said that Dr. Knellecken had been accused of the misappropriation during the three years of his presidency of DM 22 million of KZV funds, by spending it on the dissemination of educational material about the damage to health caused by the use of refined sugar. The verdict of the court rehabilitated Dr. Knellecken completely. He had been very careful to act only after obtaining the consent of his colleagues, particularly where expenses were involved, and the court found nothing that would point to undue pressure in his suggestions of the course of action to be taken by the association. *Naturarzt* added the following comment:

> Dr. Knellecken has been subject without any justification to persistent mud-slinging because of his fight for the health and well-being of patients and against the attacks on the integrity of their dentists. Dr. Knellecken, his friends and his family were publicly abused and humiliated. His position,

after thirty years of professional activity, has been gravely endangered.

Meanwhile, before this judgment had totally exonerated Dr. Knellecken, his successor in the KZV had been persuaded to sign an undertaking that any statements that KZV would make in its health promotion would in future be agreed with the sugar industry.

Nought out of ten for tact

Since most of our food comes to us as the product of some sort of agricultural activity, and since the food we eat has such an important influence on our health, it is surprising that there is so little discussion about the relationship between agriculture and nutrition. That is why I was pleased when, in June 1978, I heard that the Institute of Biology—of which I am a Fellow—had arranged a joint meeting with the Centre for Agricultural Strategy. The meeting was to consider the possible impact on agriculture that would occur if people were persuaded for nutritional reasons to reduce their consumption of milk or sugar, or change the quantities and kinds of fat they eat, or increase their consumption of dietary fiber from cereals, fruits, and vegetables.

Each of these four subjects was to be considered initially by a small panel of experts who would meet a few times before preparing a report to be presented at the Symposium in November. I was asked to chair the panel that was to consider sugar and other sweeteners.

In the middle of October the General Secretary of the Institute of Biology received a letter from which I quote:

Dear Dr. Copp,

I am writing to you in my capacity as a Fellow of the Institute rather than as Chief Executive of Tate and Lyle's Group Research and Development.

It has come as a surprise that Professor Yudkin has been chosen to speak on the general subject of "sweeteners" at the forthcoming symposium on "Food, Health and Farming," when in fact he has not done any definitive research on the subject—with the possible exception of his work on sucrose. It would have been, in my opinion, of greater interest and value to the symposium to have selected a speaker on this subject who could, at the very least, have been expected to be objective. Professor Yudkin, as you know, has in the past used symposia of this type for attacks on sugar irrespective of medical evidence which contradicts his views . . . It is indeed a pity that you have not included someone on your programme who . . . could have presented new data rather than the "same old story" which we have heard from Professor Yudkin periodically.

The General Secretary of the Institute of Biology replied to this in a letter which included the following:

Thank you for your letter of 11 October. However, I think you may not have seen the programme for the conference on "Food, Health and Farming" and so I enclose one. You will see that Professor Yudkin is presenting the report of a panel. He will therefore be putting forward views agreed by a group of responsible scientists including the Research Director of Beechams Limited.

Friendly intervention

By the early 1960s the Nutrition Department of Queen Elizabeth College had become grossly overcrowded, and the College decided that it must be extended. An appeal was launched to collect funds for this, and the then College Treasurer, who was very much connected with the food industry, wrote to his friends and acquaintances in some of

the major food companies. However, unlike all the other food manufacturers that were approached, Tate & Lyle declined the invitation to
make a contribution. The letter from the company said that the board
had given a lot of thought to the College's appeal, and continued: "You
will readily understand our Board's reluctance to support an estab-
lishment where the Professor of Nutrition considers sugar to be an
unessential item of our diet and presumably teaches this theory." The
part I like best is the last few words: I take it to mean that we might
have been given support if only I were teaching my students what I
did *not* myself believe.

During 1966 I was asked to join a small group of German doctors and dentists to meet representatives of the South German
sugar industry in a roundtable discussion of our differences. This
was, I thought, a very welcome move—better than a continuing
shouting match that clearly made no progress toward mutual understanding. We had a useful discussion; without having persuaded
the representatives of the refiners or the manufacturers that sugar
was certainly harmful, we did, I believe, convince them that we had
some justification for our concern about the effects on health.

On my return to London I wrote a letter to the then Chairman
of Tate & Lyle, describing the meeting we had had and suggesting
that this should be the pattern for our future relationship.

The Chairman replied and said he thought it would be a good
idea if I met a representative from the company. In due course a
meeting was arranged and the representative came to see me in my
office at Queen Elizabeth College. I began to talk to him about our
research, and how our new results, not yet published, were making
us even more convinced of the dangers of sugar consumption. It
soon transpired, however, that the Chairman had sent someone to
see me who was not familiar with our work. He was, in fact, the
General Sales Manager in charge of the Technical Sales Department of the company.

This was quite different from what I had encountered in Germany. And it was also an end to my hopes that I could establish a
useful dialog with people in the sugar industry.

A preemptive strike

Pure, White, and Deadly was first published in Britain in June 1972, but appeared in the United States a few weeks earlier under the title *Sweet and Dangerous*. The American publishers had taken the view that it would be useful to give a list of the then 30 or so articles in scientific and medical journals that described the experiments we had done on the effects of sugar, and set out their results. This would allow any scientist who was interested to see whether the statements in the book were justified by the results of our experiments. The publisher of the British edition, on the other hand, thought that none of its readers would be interested in such a list, and so it had been omitted.

The earlier publication of *Sweet and Dangerous* forewarned the British sugar industry of the imminent appearance in the UK of *Pure, White, and Deadly*. The then British Sugar Bureau (now the Sugar Bureau), which is the publicity arm of "British sugar refining and manufacturing," took the opportunity of producing a "News Bulletin" which they sent to newspapers, magazines and radio, and television stations that might be reviewing *Pure, White, and Deadly*. I quote just two or three items from this document:

> In this book Dr. Yudkin attributes the increase of a number of diseases mainly to the role of sugar in the modern diet.
>
> The Bureau is concerned by . . . the irresponsible way in which the evidence is presented.
>
> The book is considered to be not only unscientific in its approach, but to contain very little more than a number of emotional assertions based on Dr. Yudkin's own theory that sugar is the main cause of many diseases and should be banned.
>
> It may be significant that in the American version of this book, entitled *Sweet and Dangerous* . . . Dr. Yudkin relied for support on a selected bibliography containing a

number of references to scientific papers, nearly all of which were by Yudkin or Yudkin *et al.* In the English version of the book, however, there are no references, not even to his own published papers, to support his assertions.

I don't think it is common for a book to be publicly attacked before it is even published or reviewed.

You may think that my own few experiences illustrate a fairly restrained reaction of the sugar industry in protecting itself from what it considers unwarranted attacks on its product. If so, you will be interested to know that this is now changing; no longer will the industry respond so meekly to those who have hitherto assailed it so unjustly.

The Editor of a magazine in which I had written briefly of some of the ill effects produced by sugar received a letter in which each of my comments was vigorously criticized. The writer of the letter making these rather technical points was the Executive Director, Marketing and Sales, of British Sugar, which is the company concerned with the production and refining of beet sugar. This recalls the qualifications of the man from Tate & Lyle who came to see me at Queen Elizabeth College nearly 20 years earlier to discuss our research. Having dealt with the biochemical and clinical matters relating to my article, the letter from British Sugar proceeds, "The sugar industry now recognizes its mistake in not effectively offsetting over the years the barrage of misinformation and disinformation fostered by individuals with a desire to profit from the credulity of the population. This is in the process of being corrected."

I pointed out earlier that by no means every scientist agrees with my views about sugar. And there is nothing wrong with that: much of the material about which I have written still adds up to circumstantial evidence rather than absolute proof. But circumstantial though it is, it has been steadily accumulating over the past 20 years from several laboratories, and there is a growing number of people who believe that the case is now pretty strong that sugar is, for example, one of the causes of coronary disease.

Scientist versus scientist

I have mentioned Dr. Ancel Keys and his pioneer work in relation to diet and heart disease. In 1970 he wrote a memorandum which he sent to a large number of scientists working in this field, and which with very few changes has been published in a medical journal, *Atherosclerosis*. It consists entirely of a strong criticism of the work I have published from time to time on the theory that sugar is the main dietary factor involved in causing heart disease.

The publication contains a number of quite incorrect and unjustified statements; for instance: that we had never tested our method for measuring sugar intake; that no one eats the amounts of sugar that we and others have used in our experiments; that it was absurd of me in 1957 to use international statistics of 41 countries as evidence for the relationship between sugar and heart disease (exactly the same statistics that Dr. Keys had previously used for only six selected countries to show the relationship between fat and heart disease).

He ends by triumphantly pointing out that both sugar and fat intakes are related to heart disease, but that the cause must be fat, not sugar, because he had found in 1970 that fat intake and sugar intake are themselves closely linked. You will remember my own discussion of this point based on the fact that, as far back as 1964, I had demonstrated this same relationship between fat intake and sugar intake.

Dr. Keys has at least been consistent in his views. A rather different example of strong disagreement with our findings is provided by Professor Vincent Marks, a biochemist at Surrey University. Professor Marks and a colleague reported in the *Lancet* in 1977 some experiments which showed that taking gin and tonic could provoke hypoglycemia if the tonic water contained sugar, but not if it contained saccharin. This work was vigorously criticized in a letter to the *Lancet* by the then Director-General of the International Sugar Research Foundation—the precursor of the WSRO. Professor Marks began his reply:

May I suggest that a clue to the reason for Mr. Hugil's vitri-
olic comments on our work is to be found in his address?
The International Sugar Research Foundation must feel
threatened by the accumulating evidence that John Yudkin's description of their major product as pure, white, and
deadly is not too far wide of the mark.

By 1985 Professor Marks found himself able to write, in relation to the suggestion that sugar might be a cause of coronary disease, "one of the most groundless theories puts sugar as the villain
of the piece and is nothing more than scientific fraud." And he goes
on to say that other statements by "usually ill-informed authors
suggesting that sugar is a primary or indeed even a contributory
cause of coronary heart disease are not only false and misleading
but frankly mischievous." This remark appeared in 1985 in a color
supplement—inserted in the trade magazine the *Grocer*—written,
designed, and produced by the public relations firm working for
the Sugar Bureau.

Three months later Professor Marks was the billed speaker at
one of the "discussion meetings" organized by "Diet and Health"
which were sponsored by the Sugar Bureau. The published summary of his talk, circularized before the meeting, begins as follows:

The Diet Scandal—or are we being conned? What has
brought about the change in the public image of sugar from
that of an important constituent of the diet to that of an
unnecessary food additive responsible to a greater or lesser
extent for the host of illnesses and social miscreancies? Is it
a wealth of newly produced experimental evidence? . . . Or
is it a sensationalist bandwagon based upon nothing more
than anecdotal, incorrectly interpreted data?

I should make it clear that I have no general quarrel with scientists who change their minds. Any scientist may have to do this in
the light of new discoveries. These may show that previously held
views were based on experiments in which faulty techniques were

used, or that new observations or techniques have revealed facts hitherto unknown; in either case it will be necessary to modify the conclusions drawn from the previous observations. As far as I can see, for the period between the earlier and later opinions of Professor Marks, neither of these conditions applies. More recent experimental research carried out on the subject of sugar and disease, in several independent laboratories, has both confirmed our previous conclusions and added new observations that support them. I find it surprising, then, that Professor Marks now chooses to exculpate sugar from the accusation that it is harmful to our health. Unfortunately, such statements provide a strong and continuing source of ammunition to the sugar industry not only in defending itself but in attacking the scientists and health workers who are trying to inform the public of the need to reduce their sugar consumption.

Write what you like but only if I like it too

I suppose people mostly do not hear about efforts made to interfere with what they are doing, if these go on behind their backs. But occasionally they come to light. I was once asked to produce a slimming diet for the National Dairy Council. This project appealed to me because a sensible slimming plan must not only cut down on the total amount of food, but must do so without excessively cutting down on the essential nutrients in the food—the protein, vitamins, and mineral elements. So you aim to cut down the foods that provide little or nothing except calories and keep the foods that provide lots of nutrients in proportion to their calories. The only food that contains nothing but calories is sugar; the food with the greatest number and quantities of nutrients for its calories is milk.

This was the simple basis of the diet that I designed for the National Dairy Council. After they had published the diet, the Council was asked by one of the major sugar companies—ever so politely—to remove or at least "de-emphasize" the need to cut out sugar. The Director of the Council told me of this request, clearly expecting me to refuse to make any change; when I did refuse, he said with a smile that he fully supported me.

Pure, white—and powerful

Let me end this personal tale by repeating that I do not accuse those scientists who express disagreement with my views of doing so for improper motives. Nevertheless, I find it remarkable that there are still so many in this category, after several years of accumulating evidence that supports the conclusions that I and a few other research workers have reached. It is especially interesting that some of those who began by leaning toward accepting these views now reject them.

It is difficult to avoid the conclusion that this is the result of the vigorous, continuing, and expanding activities of the sugar interests. Their product is pure and white; it would be difficult to use these adjectives for the behavior of the producers and distributors and their intermediaries. Nevertheless, it would not be rewarding to search for an organized dirty tricks department; it seems to be more an instinctive protective action of those in the trade to deny any cover-up of the ills produced by their product, or any wrongdoing of their fraternity. The result is such a compact nucleus of power that, like a magnet surrounded by a strong induction coil, it produces a field of influence that invisibly affects many of those not in direct contact with the center.

References

2

Bothwell, D. and P., *Food in Antiquity*, Thames & Hudson (London, 1969)

Commonwealth Institute, *The Human Story* (London, 1985)

Dart, R., *Beyond Antiquity: A Series of Radio Lectures on the Origin of Man*, South African Broadcasting Co. (Johannesburg, 1965)

Leakey, R. E., and R. Lewin, *Origins*, Macdonald & Jane's (London, 1977)

Le Gros Clark, W. E., *History of the Primates*, University of Chicago Press (Chicago, 1966)

Yudkin, J. (ed.), *Diet of Man: Needs and Wants*, Applied Science Publishers (London, 1978)

3

Yudkin, J., *This Slimming Business*, 4th edn, Penguin (Harmondsworth, 1974)

Yudkin, J., *The Penguin Encyclopaedia of Nutrition*, Penguin (Harmondsworth, 1985)

4

Aykroyd, W. R., *The Story of Sugar*, Quadrangle (Chicago, 1967)

Deer, N., *The History of Sugar*, Chapman & Hall (London, 1949)

Fairrie, G., *Sugar*, Fairrie & Co. (Liverpool, 1925)

Geerdes, T., *Zucker*, Stuttgarter Verlagskontor (Stuttgart, 1963)

Geerligs, H. C. P., *The World's Cane Sugar Industry*, Norman Rodger (Altrincham, 1912)

Mintz, S. W., *Sweetness and Power*, Viking (Harmondsworth, 1985)

Strong, L. A. G., *The Story of Sugar*, Weidenfeld & Nicolson (London, 1954)

216   **References**

Eisa, O. A., and J. Yudkin, "Some nutritional properties of unrefined sugar and its promotion of the survival of new-born rats," *British Journal of Nutrition*, 54, 593 (1985)

6

Bruker, M. O., *Zucker und Gesundheit*, Verlag Schwabe (Bad Homburg, 1965)

Cleave, T. L., G. D. Campbell and N. S. Painter, *Diabetes, Coronary Thrombosis and the Saccharine Disease*, 2nd edition, John Wright (Bristol, 1969)

Yudkin, J., *The Penguin Encyclopaedia of Nutrition*, Penguin (Harmondsworth, 1985)

7

Blenford, D., "Sweetness," *Food, Flavourings, Ingredients, Processing and Packaging*, 6, 22 (September 1984)

8

Annual Abstract of Statistics, HMSO (London)

Reports of the Commission of the European Communities, Sugar Division

Friend, B., "Nutrients in United States food supply—a review of trends," *American Journal of Clinical Nutrition*, 20, 907 (1967)

Hodges, R. E., and W. A. Krehl, "Nutritional status of teenagers in Iowa," *American Journal of Clinical Nutrition*, 17 (1965)

Soft Drinks Trade Journal (London)

Sugar Year Book, International Sugar Organization (London)

Viton, A., and F. Pignalosa, *Trends and Forces of World Sugar Consumption*, FAO Commodity Bulletin Series, no. 32 (1961)

Yudkin, J., "Sugar as a food: an historical survey" and "Sucrose in the aetiology of coronary thrombosis and other diseases," *Sugar*, ed. J. Yudkin, J. Edelman, and L. Hough, Butterworth (London, 1971)

10

Yudkin, J., "Nutrition and palatability with special reference to obesity, myocardial infarction, and other diseases of civilisation," *Lancet, I*, 1335 (1963)

11

Household Food Consumption: 1983, Annual Report of the National Food Survey Committee, HMSO (London, 1985)

Orr, J. B., *Food, Health and Income*, Macmillan (London, 1937)

12

Dayton, S., M. L. Pearce, and S. Hashimoto, "A controlled clinical trial of a diet high in unsaturated fats in preventing complications of atherosclerosis," *Circulation*, 40, 1 (1969)

Lipid Research Clinics Program, "The Lipid Research Clinics Coronary Primary Prevention Trial Results: 1. Reduction in incidence of coronary heart disease," *Journal of the American Medical Association*, 251, 351 (1984)

Multiple Risk Factor Intervention Trial Research Group, "Risk factor changes and mortality results," *Journal of the American Medical Association*, 248, 1465 (1982)

Turpeinen, O., et al., "Dietary prevention of coronary heart disease: the Finnish mental hospital study," *International Journal of Epidemiology*, 8, 99 (1979)

13

McKeige, P. M., et al., "Diet and risk factors for coronary heart disease in Asians in northwest London," *Lancet, II*, 1086 (1985)

Yudkin, J., "Diet and coronary thrombosis: hypothesis and fact," *Lancet, II*, 155 (1957)

Yudkin, J., and J. Morland, "Sugar intake and myocardial infarction," *American Journal of Clinical Nutrition*, 20, 503 (1967)

Yudkin, J., and J. Roddy, "Levels of dietary sucrose in patients with occlusive atherosclerotic disease," *Lancet, II*, 6 (1964)

14

Ahrens, R. A., "Sucrose, hypertension and heart disease: an historical perspective," *American Journal of Clinical Nutrition*, 27, 403 (1974)

Bruckdorfer, K. R., I. H. Khan, and J. Yudkin, "Dietary carbohydrate and fatty acid synthetase activity in rat liver and adipose tissue," *Biochemical Journals*, 123, 7 (1971)

Irwin, M. I., and A. J. Staton, "Dietary wheat starch and sucrose: effect on levels of five enzymes in blood serum of young adults," *American Journal of Clinical Nutrition*, 22, 701 (1969)

Reiser, S., et al., "Isocaloric exchange of dietary starch and sucrose in humans: 1. Effects on levels of fasting blood lipids," *American Journal of Clinical Nutrition*, 32, 2195 (1979)

Reiser, S., et al., "Isocaloric exchange of dietary starch and sucrose in humans: 11. Effect on fasting blood insulin, glucose, and glucagon and on insulin and glucose response to a sucrose load," *American Journal of Clinical Nutrition*, 32, 2206 (1979)

Reiser, S., M. Blickerd, J. Hallrisch, O. E. Michaelis, and E. S. Prather, "Serum insulin and glucose in hyperinsulinemic subjects fed three different levels of sucrose," *American Journal of Clinical Nutrition*, 34, 2348 (1981)

St. Clair, R. W., B. D. Bullock, N. D. M. Lehner, T. B. Clarkson, and H. B. Lofland, Jnr, "Long-term effects of dietary sucrose and starch on serum lipids and atherosclerosis in miniature swine," *Experimental and Molecular Pathology*, 15, 21 (1971)

Samsonow, M. A., and W. A. Meschtscherjakowa, "Der Einfluss qualitativunterschiedlicher Kohlenhydrate auf den Verlauf der Koronaratherosklerose," *Ernährungsforschung*, 13, 331 (1968)

Szanto, S., and J. Yudkin, "The effect of dietary sucrose on blood lipids, serum insulin, platelet adhesiveness and body weight in human volunteers," *Postgraduate Medical Journal*, 45, 602 (1969)

Szanto, S., and J. Yudkin, "Dietary sucrose and platelet behaviour," *Nature*, 225, 467 (1970)

Taschev, T., and G. Patschewa, "Über die Bedeutung der Kohlenhydrate bei der Ätiopathogense der experimentellen Atherosklerose," *Ernährungsforschung*, 13, 339 (1968)

Vrána, A., and L. Kazdová, "Insulin sensitivity of rat adipose tissue and of diaphragm in vitro: effect of the type of dietary carbohydrate (starch—sucrose)," *Life Sciences*, 9, 257 (1970)

Yudkin, J., and R. Krauss, "Dietary starch, dietary sucrose and hepatic pyruvate kinase in rats," *Nature*, 215, 75 (1967)

Yudkin, J., S. Szanto, and V. V. Kakker, "Sugar intake, serum insulin and platelet adhesiveness in men with and without peripheral vascular disease," *Postgraduate Medical Journal*, 45, 608 (1969)

Yudkin, J., and S. Szanto, "Hyperinsulinism and atherogenesis," *British Medical Journal*, 1, 349 (1971)

15

Cohen, A. M., S. Bavly, and R. Poznanski, "Change of diet of Yemenite Jews in relation to diabetes and ischaemic heart disease," *Lancet*, II, 1399 (1961)

Cohen, A. M., H. Freund, and E. Auerbach, "Electroretinogram in sucrose-and starch-fed rats," *Metabolism*, 19, 1064 (1970)

Cohen, A. M., and A. Teitelbaum, "Effect of dietary sucrose and starch on oral glucose tolerance and insulin-like activity," *American Journal of Physiology*, 206, 105 (1964)

Cohen, A. M., and J. Yudkin, "The effect of dietary sucrose upon the response to sodium tolbutamide in the rat," *Biochimica et Biophysica Acta*, 141, 637 (1967)

Kang, S. S., R. G. Price, K. R. Bruckdorfer, N. A. Worcester, and J. Yudkin, "Dietary induced renal damage in the rat," *Proceedings of the Nutrition Society*, 36, 27A (1976)

Kang, S. S., R. G. Price, J. Yudkin, N. A. Worcester, and K. R. Bruckdorfer, "The influence of dietary carbohydrate and fat on kidney calcification and the urinary excretion of N-acetyl-ß-glucosaminidase (EC 3.2.1.30)," *British Journal of Nutrition*, 41, 65 (1979)

Mather, H. M., and H. Keen, "The Southall diabetes survey: prevalence of known diabetes in Asians and Europeans," *British Medical Journal*, 2, 1081 (1985)

Papachristodoulou, D., H. Heath, and S. S. Kang, "The development of retinopathy in sucrose-fed and streptozotocin-diabetic rats," *Diabetologie*, 12, 367 (1976)

Price, R. G., S. A. Taylor, S. S. Kang, K. R. Bruckdorfer, and J. Yudkin, "Composition and biosynthesis of rat glomerular basement membrane in sucrose-fed rats," *Glycoconjugate Research*, 2, 747 (1979)

Rao, P. N., V. Prendiville, A. Buxton, D. G. Moss, and N. J. Blacklock, "Dietary management of urinary risk factors in renal stone formers," *British Journal of Urology*, 54, 578 (1982)

Rosenmann, E., A. Teitelbaum, and A. M. Cohen, "Nephropathy in sucrose-fed rats," *Diabetes*, 20, 803 (1971)

16

Gass, D. J., *Curing Ulcer Disease*, Pioneer Publishing Co., Fresno, California, 1983

Heaton, K. W., "The role of diet in the aetiology of cholelithiasis," *Nutrition Abstracts and Reviews* 54, 549 (1984)

Scragg, R. K. R., A. J. McMichael, and P. A. Baghurst, "Diet, alcohol, and relative weight in gall stone disease: a case-control study," *British Medical Journal*, 1, 1113 (1984)

Scragg, R. K. R., G. D. Calvert, and J. R. Oliver, "Plasma Lipids and insulin in gall stone disease: a case-control study," *British Medical Journal*, 2, 521 (1984)

Thornton, J. R., P. M. Emmett, and K. W. Heaton, "Diet and Crohn's disease: characteristics of the pre-illness diet," *British Medical Journal*, 2, 762 (1979)

Thornton, J. R., P. M. Emmett, and K. W. Heaton, "Smoking, sugar and inflammatory bowel disease," *British Medical Journal*, 1, 1786 (1985)

Werner, D., P. M. Emmett, and K. W. Heaton, "Effects of dietary sucrose on factors influencing cholesterol gall stone formation," *Gut*, 25, 269 (1984)

Yudkin, J., E. Evans, and M. G. M. Smith, "The low carbohydrate diet in the treatment of chronic dyspepsia," *Proceedings of the Nutrition Society*, 31, 12A (1971)

17

Dental caries

Bowen, W. H., B. Cohen, M. F. Cole, and G. Colman, "Immunisation against dental caries," *British Dental Journal*, 139, 45 (1975)

Children's Dental Health 1983, Office of Population Censuses and Surveys, HMSO (London, 1983)

Jenkins, N., "Diet and dental caries," *Food and Nutrition News*, 56, 29 (1984)

Salter, A. J., and J. Yudkin, "Dental caries and between-meal snacks" (letter), *British Medical Journal*, 1, 577 (1978)

Sheiham, A., "Changing trends in dental caries," *International Journal of Epidemiology*, 13, 142 (1984)

Todd, J. E., A. M. Walker, and P. Dodd, *Adult Dental Health, UK*, Office of Population Censuses and Surveys, HMSO (London, 1978)

Skin disease

Bett, D. G. G., J. Morland, and J. Yudkin, "Sugar consumption in acne vulgaris and seborrhoeic dermatitis," *British Medical Journal*, 3, 153 (1967)

Disease of the liver

Bender, A. E., K. B. Damji, M. A. Khan, I. H. Khan, L. McGregor, and J. Yudkin, "Sucrose induction of hepatic hyperplasia in the rat," *Nature*, 238, 461 (1972)

Best, C. H., W. Stanley Hartroft, C. C. Lucas, and J. H. Ridout, "Liver damage produced by feeding alcohol or sugar and its prevention by choline," *British Medical Journal*, 2, 1001 (1949)

Cancer

Bristol, J. B., P. M. Emmett, K. W. Heaton, and R. C. N. Williamson, "Sugar, fat and the risk of colorectal cancer," *British Medical Journal*, 2, 1467 (1984)

Seely, S., and D. Horobin, "Diet and breast cancer: the possible connection with sugar consumption," *Medical Hypotheses*, 11, 319 (1983)

Drug action

Nash, A. H., and A. E. Bender, "The effect of dietary sucrose on the metabolism of pentobarbitone," *Proceedings of the Nutrition Society*, 35, 132A (1976)

18

Dalderup, L. M., and W. Visser, "Influence of extra sucrose in the daily food on the life-span of Wistar albino rats," *Nature*, 222, 1050 (1969)

Taylor, D. D., et al., "Influence of dietary carbohydrate on liver lipid content and on serum lipid in rélation to age and strain of rat," *Journal of Nutrition*, 91, 275 (1967)

Ziegler, E., "Die Ursache der Akzeleration, ernährungsphysiologische und medizinhistorische Betrachtungen über den Zuckerkonsum des modernen Menschen," *Helvetica paediatrica Acta*, 21, supple. 15, 1 (1966)

19

Yudkin, J., "Dietary factors in arteriosclerosis: sucrose," *Lipids*, 13, 370 (1978)

21

Imfeld, A., *Zucker*, Unionsverlag (Zurich, 1983)

Index